PETER D.

Peter worked at the Royal Ontario Museum for over thirty years, traveling the world collecting specimens for the institution. He worked in several disciplines in the museum, from the science departments to head of both the outreach and the exhibit design.

For Joseph and Bronwyn.

Peter Buerschaper

LIFE AT A MUSEUM

AUSTIN MACAULEY PUBLISHERS™

LONDON • CAMBRIDGE • NEW YORK • SHARJAH

Ordering Information:
Quantity sales: special discounts are available on quantity purchases by corporations, associations, and others. For details, contact the publisher at the address below.

Publisher's Cataloging-in-Publication data
Buerschaper, Peter
Life at a Museum

ISBN 9781643781228 (Paperback)
ISBN 9781643781235 (Hardback)
ISBN 9781645366966 (ePub e-book)

Library of Congress Control Number: 2019908479

The main category of the book — BIOGRAPHY & AUTOBIOGRAPHY / Adventurers & Explorers

www.austinmacauley.com/us

First Published (2019)
Austin Macauley Publishers LLC
40 Wall Street, 28th Floor
New York, NY 10005
USA

mail-usa@austinmacauley.com
+1(646)5125767

Thanks to Margaret and Mark Buerschaper.

Prologue

The following is more or less a summary of my life at the Royal Ontario Museum (ROM). I have tried to include most of my experiences working in this institution – both good and bad. I still think highly of the ROM and believe that the museum has great potential. The latter has, at times, not been realized due to staff shortcomings. The ROM is a great research institution and a wonderful educational experience for the general public.

Ichthys and Herps, and Beyond

The Royal Ontario Museum (ROM) opened on March 19, 1914. The first museum building was built along the University of Toronto's Philosopher's Walk with its entrance off Bloor Street. Originally, the institution was made up of five separate museums which included the museums of archaeology, paleontology, mineralogy, zoology, and geology.

Since its opening, the ROM has always been the largest museum in Canada and over the years, numerous wings for public displays and for curatorial research facilities have been added making the ROM quite a substantial institution.

It was indeed a sad day when I left the Royal Ontario Museum (ROM) for the last time. It was a cool, dingy gray October day in 1991 when I packed up my personal belongings and drove home leaving the only workplace I ever really knew. I informed the head of human resources that I was resigning. He, like some little boy, immediately ran upstairs to the director's office to inform him that he got another one. This head of human resources, I think, was hired as a headhunter. Unfortunately, at that time, this was the kind of institution it had become. Like much within the overall museum, the department in which I worked had changed dramatically. For the most part, I always supported change when it was in a constructive direction. Senior management in its wisdom had set up a working structure I found to be most difficult. With all of the infighting, office politics, union interference, and overall negative working conditions, it had become practically impossible to accomplish anything satisfactory in the exhibit design area. Creativity was out the window. Everybody was in charge of everything and originality was lost in the competitive attitudes displayed by the many self-serving individuals who had been added to the department.

I suspect that my experiences at the ROM exist in many companies/institutions, and are for the most part human nature.

My first introduction to the ROM was when I was twelve years old – on a school outing. I had just arrived in Canada from Europe and was now a resident of Toronto. We lived in a small semi-detached house, at 33 LeMay Road, located in the Bayview and Millwood area. My dad had bought this dwelling just before our arrival in our new country in 1953. Dad came to Canada, with one dollar in his pocket, by himself a year before. He worked two jobs. As butcher during the day and in the evening as a chef in an upscale restaurant, the latter being his profession – master chef. He thus provided a great residence for my mother, my younger sister, and myself. All was new to me. Having lived in a small town in the old country, Bayview Avenue seemed gigantic. Stores like Kresge's and Loblaws appeared huge and self-service stores were totally foreign to me. I had many problems trying to buy groceries for my mother, not knowing the names of any items.

The person who initially introduced me to the ROM was my first Canadian school teacher – Mr. Hunt. Apparently, he was a retired RCMP officer, an average sized man with a gentle appearance. Mr. Hunt taught several grades of only boys in a one-room portable schoolhouse which, in the winter time, was heated by an old fashioned coal stove with a long stove pipe disappearing through the high ceiling of the classroom. The walls of this rather unique classroom were decorated with large maps describing the various provinces and territories of Canada. At the time, I found it odd that these maps were decorated with pictures of chocolate bars; however, I soon realized that certain chocolate companies sponsored these colorful maps promoting their products. My new classroom was located on the grounds of Maurice Cody Public School.

Each day after school, I walked Bayview Avenue to explore my new world. Without any command of the English language, it took some time to get to know the names of everyday common items. I had never experienced such large stores, where you helped yourself to whatever you wanted to buy. I would often stop at a television store, peering through its window to watch in silence any given program. I thought this to be quite magical, having never experienced TV before. All these totally new encounters were challenging, yet exciting.

Mr. Hunt was a wonderful person. I will always be grateful to him for his continuous encouragement and thoughtful guidance to get me started in my new life. At one time, I'm not sure how we managed to communicate, since my command of the English language was still pretty much non-existent. I'm not sure how I told Mr. Hunt that I was raising guppies at home in one of my fish tanks. He told me that he too had an aquarium at home and that he would like to have some of these for his fish tank. I brought to him a dozen or so of my fanciest guppies which made him very happy and he very much appreciated my gesture. At that time, he also found out, and I still don't know how, that I had an interest in art and he soon taught me how to paint, in transparent watercolors, a Christmas flower in various rich shades of brilliant reds. Mr. Hunt must have done some painting in his spare time because he sure knew what he was doing. I am sorry that in later years I never managed to get in touch with him to personally thank him for all he had done for me. I guess I was too busy establishing myself in my new world.

I guess schoolteachers in those days did not make a lot of money as Mr. Hunt pumped gas at an Esso station on Front Street near the CNE grounds on weekends and during his summer vacation. Although everything has changed over the years, the gas station is still there.

When Mr. Hunt took all of his students to the Royal Ontario Museum, my command of the English language was still pretty much non-existent. Yet that first visit to the ROM, where I got to marvel at the displays, left a long-lasting impression on me – especially those exhibits involving natural history. I also remember being impressed by the North American native gallery in the basement of the museum. Little did I know, at the time, that these replicas had all been modeled from white men and did not visually represent native people.

Following that first visit, I returned to the ROM many more times, mainly on weekends, whenever I wanted to relive some of my original experiences. Even then, at that early age, I often asked myself how would I get a job at this institution and how would I go about doing so?

I should mention at this time that after leaving school and before my work introduction at the ROM, I served a two-year apprenticeship as an electrician. This was quite customary in the old country when having finished public school, one would enter an apprenticeship to learn a trade. This I did not like at all and decided to go back to school at which time I was introduced to do some work for the Royal Ontario Museum.

At the age of eighteen, when through my continuing interests in fish and aquaria, I met a ROM staff member by the name of E. H. Taylor who at that time was in charge of the ROM zoological laboratory. There he prepared zoological specimens for all the ROM collections for the various life science departments. One of the highlights in the zoo laboratory was a bathtub, standing against its southern wall, in which a three-foot live alligator made its home. The gator was fed the skinned-out bodies of mammals and birds which had been made into study skins for the respective life science departments. Mr. Taylor was also in charge of the fish tanks displaying live native fish in the third-floor rotunda of the public display areas. He hired me to maintain these fish tanks and while attending high school, I was able to make some pocket money and I got to know many other museum staff members finding my way around several curatorial departments. During this part-time stay at the museum, I was introduced to many museum curators, one of which was the original curator of herpetology – Shelley Logier. Unfortunately, Shelley retired shortly after my arrival. Not only was Shelley a very knowledgeable herpetologist but was also a terrific artist and illustrator. He wrote and published several major books on Ontario reptiles and amphibians, each of which he illustrated mostly in transparent watercolors. I always remember examining one of his original watercolors of a life-sized Ontario bullfrog which was absolutely gorgeous.

Shelley always kept an aquarium, covered with a securely fitted metal screen lid, full of some massasauga rattlesnakes. If anyone wanted to see any of these snakes close up, Shelley simply removed the screen top off the tank and, in a swift move, reached into the tank and grabbed, between his thumb and forefinger, directly behind a chosen snake's head and brought it forth to show it to any interested visitor. That was one trick of Shelley's I had no interest in repeating. It was not until later on during my stay at the ROM that I realized how pleasurable it was to have met one of the original curators of the institution.

After two years of maintaining the ROM's live fish displays, I was informed by Mr. Taylor that he had accepted a position in Calgary Alberta. He was to become the director of a newly built Public Aquarium. He asked me to come along to help him to get the project off the ground. I accepted and spent two and a half years at the Calgary Aquarium setting up large display tanks holding various live native and exotic freshwater and saltwater specimens. This included a large saltwater tank displaying several mature green sea turtles. These animals required a varied diet ensuring their health and comfort in their artificial habitat.

From time to time, I accompanied El Taylor to the west coast, to Vancouver, British Columbia. There, with assistance of the Vancouver Aquarium staff, we collected a variety of Pacific Ocean fish which we would take back to Calgary to display in the aquarium building. In Vancouver, we usually set out on the Vancouver Aquarium's collecting boat out of the Horseshoe Bay Marina and with a small trawl, collected a variety of Pacific Ocean fish for the Calgary Aquarium. We caught a variety of Pacific rockfish and some invertebrates. A couple of times we managed to collect a large Pacific octopus. Displaying it at the Calgary Aquarium always attracted many more visitors than the usual crowd.

Once in a while, we caught a net full of Dungeness crabs which we boiled in salt water and eagerly consumed with melted butter on a nearby picnic table in Horseshoe Bay. A rare treat indeed!

During my stay in Calgary, once in a while I took time to drive home for short visits to Toronto. On one of these occasions, on my return trip to Calgary, George McLean (a superb wildlife artist) joined me and came along to check out a couple of western sites in which he was interested. On our trip West, we drove through some substantial rain and when on one occasion I turned on the heating system in my old Volkswagen, all we got was cold water spraying out of the heat vents.

George wanted to visit the Edmonton game farm and he wanted to view the art collection at the Glenbow institution in Calgary. We did visit both sites and thoroughly enjoyed both venues. We experienced close-up interaction with numerous native animals at the game farm. One that comes to mind was hand feeding a wolverine, who, with some hesitation, took pieces of raw meat from our hands. At the Glenbow, I was introduced to the work of the self-taught cowboy artist, Charlie Russell. When viewing his paintings, I immediately appreciated his use of colors, dramatic compositions, and anatomical accuracy. His technique made the work stand out uniquely from his contemporaries. Charlie told stories of the Old West in his works. Many of which dramatized the past lives of Native North American peoples. He also personally associated himself with many of the stories he told. His descriptive titles not only added humor but a visual understanding of his drawings and paintings.

After a couple of weeks, George flew back to Toronto and my part-time duties at the Aquarium continued for some time longer. However, my employment in Calgary, too, was only part time and I soon decided that a full-time job was in order – preferably at the ROM in Toronto. I telephoned Dr. W. B. Scott, the head curator of the ROM's Department of Ichthyology and Herpetology, whom I had met during my part-time stay at that museum. Surprisingly, he told me, "Come on down, Pete, we have an opening for a technician in our department." I enthusiastically accepted the offer and almost immediately packed up my Volkswagen Bug and drove hurriedly back to Toronto. By this time, I had reached the ripe old age of twenty-one. I had become a Canadian citizen and I was ready to permanently settle into my future.

It was the middle of April in 1962 when I arrived back in the East and soon thereafter started my full-time job at the Royal Ontario Museum. When I reflect on my first day at my new full-time job, I remember that I managed to arrive late for work. I did not know the best route from Scarborough to the Royal Ontario Museum.

On that first day, it soon became apparent that I was now exposed to many unfamiliar tasks. However, it really did not take a lot of time to learn these tasks and everything soon worked out to everyone's satisfaction. Within a relatively short time, I learned the ways of the department and constantly learned more and more about what a research museum was all about.

Sometime before my arrival, Dr. W. B. Scott, affectionately known as W. B., the head curator of the department, was an average-sized man who always wore glasses. These he would push up to his forehead when examining any small fish specimen close up. Before my arrival, W. B. had written and published a book on the *Fish of Ontario*. He and Dr. E. J. Crossman, the latter was the second curator in the department, had also just completed an extensive trip to Newfoundland collecting all freshwater fish species native to the Province. This collecting expedition resulted in the publication of a book on the freshwater fish of Newfoundland. I was sorry to have missed out on that collecting trip.

During those early days in the department, all men wore white shirts and neckties, and a white lab coat for protection. Being a practical person, W. B. always wore a bow tie because it would not accidently dip into the fish preservatives when examining specimens.

W. B. was very much a hands-on sort of guy. One of his unique qualities was that he was rarely impressed by the academic statute of any given person. He respected people for what he or she could do, never paying all that much attention to their formal education. I always thought, and still do, W. B. to be an exceptional human being. His continuous genuine care and respect for other people was a truly rare and remarkable quality. He did a lot for others and helped countless people to get settled in their future professional work.

It was during my initial days at the ROM when W. B. invited me into his office, explaining a job to me that he, for some time, wanted done. The job was to expand the department's storage facilities. He told me, "This job could not be done by some outside contractor." He explained to me, "We simply don't have the money for that and I thought you might be able to help us out." I must admit it was not the sort of thing I expected to be doing when I joined the department. What was I to do – refuse? Not very likely. Anyhow, soon I was building storage boxes out of three-quarter-inch thick pre-cut sheets of plywood. These were going to be used for storing larger preserved specimens which would not fit into any collection room churns or bottles. I built these three-by-two-foot boxes complete with tightly fitting lids, and lined each with a heavy gray epoxy resin to hold in the 65% alcohol preservative and specimens. At the same time, and in the same manner, I also constructed two much larger boxes, eight feet in length and four feet in height and width. These boxes were to store the largest of the departments' preserved fish.

I felt like a classic English man, wearing a tie and white shirt while constructing storage boxes and lining these with the thick, gooey gray epoxy resin. The latter was quite a messy job. I managed to ruin more pants than I cared to admit. Fortunately, building the storage boxes was only part of my daily routine. Just the same, this construction of boxes went on for quite some time – for at least two or three years. The saving grace in all of this laborious work was that these storage boxes solved a lot of storage problems and are still in use after forty-plus years.

I was also put in charge of preparing all preservatives for all the new, fresh incoming specimens. All new unpreserved fish and reptile arrivals had to be fixed in 10% formaldehyde; then washed for a couple of days in fresh running cold water and transferred to 65% denatured alcohol for permanent storage. All new arrivals were given an accession number to record their origins. These specimens were then sorted, identified, measured, and – if needed for the collection – cataloged, and added to the three-dimensional library in the collection rooms.

The Department of Ichthyology and Herpetology occupied the entire basement of the center wing, located directly under the ROM's old exhibition hall. Although located in the basement, all offices, laboratories and the classroom had tall, out-of-the-ground windows which made working conditions quite pleasant. To get to the Department of Ichthyology and Herpetology in the old museum, you entered through a set of double steel basement doors off the museum gallery spaces. These were located opposite to the bottom of the ROM's northern visitor's staircase, and directly opposite to the bottom of the northern large totem pole. The latter rising from the basement floor, beside the staircase, up through three floors stopping just short of ROM's roof ceiling. Coming through these steel basement doors, before entering the ichthyology department, also led to the ROM guardroom and to the back entrance from the

ROM parking lot. Entering through these basement doors and after immediately turning right, you entered a long L-shaped corridor which eventually led to the Department of Ichthyology and Herpetology.

Going down the L-shaped hallway, passing by the general zoological laboratory and the 'bug-room' (a room dedicated to the cleaning of animal skeletons by domestic beetles) and then turning ninety degrees to the left, you entered the ichthyology and herpetology department. The first room of the Department, on the west side of the hallway, had one entrance but it was a room divided by an open doorway into two separate rooms. The first room was used by ROM's cleaning ladies as their home base. The second room was a storage room. Here, a cabinet with large drawers stored all maps and all original historical artworks used in various past publications.

With fondness, I can still hear the noisy vocalization and excessive laughter of the two cleaning ladies, Gladys Wong and Mrs. Tinny. Gladys was a rather buxom lady with bright red hair which was accentuated by her cobalt blue dress uniform. Gladys was a very generous lady and liked by all she encountered. Mrs. Tinny was a small slightly built lady with a no-nonsense attitude. These two ladies would often tell each other stories about the goings-on within the various ROM departments. Their often noisy and mostly cheerful discussions provided much amusement and real knowledge to all of us in the ichthyology and herpetology department about other staff members of the museum.

A bit further down the hallway, on the east side, was a small office periodically used by Dr. J. R. Dymond. Dr.

Dymond was retired from his position as the original Director of the ROM's Museum of Zoology. It is worth noting that just outside, above Dr. Diamond's office door, hung several large mounted Atlantic salmon. These had been collected, years ago, in Wilmot Creek, just east of Toronto and had been mounted in realistic poses. Atlantic salmon were once quite common in Lake Ontario waters.

Next, down the hallway, directly opposite from one another, were two large rectangular rooms. The room on the east side of the hallway ran parallel to the ROM parking lot and the other on the west side of the hallway faced U of T's philosopher's walk. Each of these rooms had tall windows providing each with plenty of daylight. The room on the west side of the hallway was the Department's classroom. This facility was complete with a substantial teaching collection of preserved fish. The classroom was used weekly, during the university teaching season, to teach ichthyology to a variety of University of Toronto zoology students.

Opposite the classroom, an equal in size room housed the herpetological collection. This room was also used as a construction area and occasionally by students to accomplish some of their work. A door in the southwest corner of this room led to the Department's secretarial office.

The east wall of the corridor between the herpetological collection/construction room and the secretarial office was lined with floor-to-ceiling closed cupboards storing the Department's collection equipment. At the end of these cupboards, on the east side of the hallway, was the official entrance to Department's secretarial office. When entering the secretarial office, immediately on the right stood a waist-high, large, flat-topped cupboard storing all the departmental tools and the top of this cupboard was used to assemble and package all outgoing shipments.

The secretarial office was large enough to accommodate two secretaries and did so in later years. A closed door in the south and west corner of the office led to the live room were all live specimens were housed. Just before entering the live room, on the right of the secretarial office was a small space/room not much larger than a closet, which was occasionally used to make coffee for the department's staff. The live room, to the south of the secretarial office, was a long, narrow room without any windows which had at one time been used as a photographic darkroom. The southern wall of the live room was lined with ten-gallon aquaria perched on a shelf some five feet off the ground. Directly underneath the aquaria, a row of fifty-gallon fiberglass tubs lined the floor from one end to the other. All of these holding tanks often housed different species of live fish used for the ROM gallery aquaria displays or for various scientific observations by the curators. The live room required daily attention – feeding its live inhabitants and from time to time draining some of the old water in each tank and replacing it with fresh water to keep the living occupants healthy.

A bit down the hallway from the secretarial office and on the opposite side was the office of Dr. W. B. Scott. It was a spacious setting with a formal old-fashioned oak desk and several chairs. A three-foot-wide shelf some thirty inches off the floor lined all the remaining office walls. These shelves provided workspace to examine specimens for any ongoing research. A sink with both hot and cold running water was inserted into the shelving directly under the middle window of the office. Above the shelves, all walls were covered with bookshelves and solidly filled with books and scientific papers. W. B. frequently had visitors, usually curators from other ROM departments. Apparently, during these visitations ROM policies, how to run and or how to improve the institution, were often discussed. I remember vividly how upon conclusion of the lengthy sessions between W. B. and a visiting curator, the office doors would finally open to reveal puffs of cigarette, cigar or pipe smoke which unfurled into the hallway.

Further down the hallway, also on the west side, was the office of Dr. E. J. Crossman. E. J. was a large man with a voice complementing his size. His office also had a large oak desk, several chairs, and a thirty-inch-wide flat top shelving lining the remaining walls. Their function was pretty much the same as those in W. B.'s office. A sink with hot and cold running water was also present under one of the office windows. E. J.'s office walls, too, were lined with shelves loaded with books and scientific papers.

The east hallway wall, opposite to the curatorial offices, was also lined with floor-to-ceiling closed cupboards. These cupboards were crowded with hundreds of jars of various sizes; filled with collections of preserved Ontario fish.

Finally, the long hallway which divided the department's workspaces ended at a pair of large metal double doors, leading to the main fish collection room. When entering this large room, two forty-five gallon drums containing 95% percent denatured alcohol and all other preservative fluids, including full-strength formaldehyde, lined the right wall of this room. This large collection room was L-shaped and did not have any windows. The room was crowded with rows of floor-to-ceiling steel shelves; separated by long, narrow walkways providing access to thousands of bottles containing thousands and thousands of alcohol-preserved fish specimens. Under all rows of shelving, on the floor, large porcelain churns and five-gallon bottles filled with larger fish specimens completed the ROM's fish collection.

Each summer, we hired a couple of high school students to clean and top up every specimen bottle in the collection room. These students added a lot of fun and humor to the Department. One day, while I was walking up the hallway and passing Dr. Scott's open office door I found him leaning back in his chair laughing heartily. I stopped and asked him, "What's so funny?"

He promptly answered, "I'm listening to our boys in the collection room." He had the intercom system on, which was connected from his office to the collection room, and he was listing to the students telling each other dirty 'Ernie' jokes. These jokes often contained vocabulary rarely heard in the museum.

When you entered the collection room and turned immediately to the right, a pair of doors led to the Department's main Laboratory. The lab was a long room, as long as the southern width of the main collection room. The lab was lined with tall out-of-the-ground windows on its south and western sides. One of the two south windows of the lab had been made into a door, leading to the parking lot. Presumably, this was to serve as a fire escape. It was, however, frequently used as an easy exit to the cars in the ROM parking lot.

In the center of the laboratory stood a large, epoxy-covered worktable. This table had a large stainless steel sink at its northern end and a long trough-like epoxy lined sink running down its center. The latter was used to wash formaldehyde-preserved specimens in fresh running water after which these were transferred into 65% alcohol for permanent storage. On a sturdy shelf, supported by the lab tabletop, some five feet off the lab floor, stood a large ten-gallon plastic tapped container dispensing the sixty-five percent alcohol used as the final specimen storage preservative.

The north wall of the lab originally had one desk and sometime later, two – used by the technicians to hand catalog all specimens for the collection. The wall separating the main collection from the lab was lined from floor-to-ceiling cupboards containing all necessary lab supplies, including a variety of lids and bottles used for final storage for all cataloged specimens.

In the winter, the old hot water radiators, heating the department, sang their unique songs when the hot water in them gurgled and squealed continuously. The museum, without air conditioning, made some hot and humid summer days challenging. On some of those hot days, you could literally see the alcohol evaporating from any open specimen bottles.

I spent the first summer pretty much by myself in the department. The secretary and both curators were on holiday and/or working in the field. I guess I was in charge of everything. I continued to build new storage boxes, cataloged specimens, answered the telephone, and pretty much attended to everything that needed to be done and I was able to do it all. It gave a great opportunity to familiarize myself with a lot of the Department's holdings, and its overall services. I remember searching through most of the cabinet drawers in the map/art storage room, next to the lady cleaners home base, to examine lots of original drawings and watercolors, most of which had been created by Shelley Logier.

I also began to empty some of the hall cupboards of many bottles of Ontario fish which had been collected by the then called 'Ontario Department of Lands and Forests.' Most of these specimens were still stored in formaldehyde and needed to be washed in running water before being transferred to sixty-five percent alcohol. I soon familiarized myself with artificial identification keys to identify many of the small Ontario fish. Soon, I managed to identify many of these Ontario fish by sight. This exercise was challenging but very satisfying. That first summer went by quickly and by the end of it, I started to feel at home in the Department of Ichthyology and Herpetology.

In September of my first year at the ROM, all of the staff returned to the department. I continued to do all of the above chores, plus whatever was requested of me by the curators. I cataloged all the newly received specimens for the collection and shipped them out. I received many loans to and from other institutions as well. I assisted in setting up the classroom with W. B. for weekly teaching sessions and retrieved lots of bottles of preserved fish requested by the curators for the Ichthyology teaching classes. During these times, I managed to have numerous discussions with both curators and soon learned a lot about ichthyology and all that was associated with the subject.

The first fall and winter of that year went by quickly. I got to know many other ROM staff members outside of our Department and thoroughly enjoyed all there was to learn from numerous seasoned experts. One of the staff I got to know fairly well was Jim Bailey – he was a curator in the ornithology department of the museum. Jim was a substantial sized man. He was at least six feet tall, had strawberry-colored hair, and had a gentle facial expression which made him very approachable. He always had a pipe in his mouth – lit or not. My first encounter with Jim was when he asked me and some other ROM staff members to help him move some large, third-floor oak and glass display cases from one area to another. Another person included in this laborious move was Tosh Yamamoto who started to work at the ROM in the entomology department about the same time I did.

The mammal display cases we moved about were large. All of these cases displayed full-sized mounted mammals and took all of our efforts and strength to slide them around the third-floor gallery space. One of the cases housed a mature, male moose with a set of full-sized antlers. During the shuffling of these display cases, I discovered Jim Bailey's sense of humor and soon got to know him well. Jim and other staff members would, from time to time, meet in the Park Plaza Pub after work to have a beer and to compare recent museum experiences. At such times we all enthusiastically listened to Jim's tales of past museum adventures and to the many jokes he insisted on telling. I guess his jokes were funny. However, while chewing on his pipe, Jim would often break into laughter making it hard to catch any of his jokes' punch lines. Not often did any of us get to understand the full meaning of his jokes. Just the same, Jim's jokes were always delightful to listen to.

Jim Bailey was one of the most accessible persons in the museum. He helped many bird artists to get established. One great example of these artists was Fenwick Lansdowne. The latter had his first one-man show in the third-floor rotunda of the ROM which was organized by Jim Bailey. Jim was always there to assist anyone interested in birds. He paid special attention to young people and really encouraged them to learn about bird identification, bird songs, and bird behavior. He often took groups of people to the east of Toronto to the Whitby marshes where they registered bird sightings. Jim informed his followers how to identify birdcalls and any sighted bird species.

Unfortunately, after a couple of years after my arrival at the ROM, Jim Bailey retired. This turned out to be a rather sad affair. By the time of his retirement, senior management had already hired a new, young head curator, from somewhere in the USA, to take his place. Jim wanted to continue to do some of his work at the ROM after retirement. He wanted to come into the ornithology department and spend some time working on certain projects that he wanted to complete. He wanted to continue his services to the public. However, the newly appointed curator would, for whatever reason, not allow Jim to come into the department to continue his lifelong passion for birds and his dedication to his work. Jim got tremendously upset and could, unfortunately, not find any support within the ROM to assist him in this matter. Jim was so upset that shortly after his retirement he had a heart attack and died. A very sad affair. Jim was truly missed by all staff and certainly by the public.

Following the sad episode of Jim Bailey, and perhaps because of it, I got to know Terence Michael Shortt – the head of the ROM's live science display department. Known as 'Terry' to most who knew him, he was one of Canada's foremost bird artists and was very knowledgeable on most museum subjects. Terry built and painted several major ROM dioramas, each designed on his field trips to various exotic places. When I first met Terry, he and his staff were just completing the design and were installing a new reptile gallery. Some of the preparations for this gallery were still ongoing. A couple of large snakes that were to be a part of this new gallery arrived at the ROM shortly after I met Terry Shortt. One of these came from India – a large king cobra. The other was a mature, full-sized anaconda from South America. The cobra was packaged in a solidly constructed wooden crate and no one knew whether or not this reptile was dead or alive. Not taking any chances, the cobra crate was immediately placed in a freezer. Since it is a cold-blooded animal, this was the most humane way to kill the snake. The large anaconda was definitely alive when it arrived. It had been shipped in a large steel drum which was secured with a tightly fitted steel mesh lid. When we decided to look through the reinforced mesh lid to more closely examine this large reptile, the wide open anaconda mouth instantly hid the underside steel mesh lid, totally discouraging me to engage in any further examination. Both of these large snakes, after death, were manipulated into life-like positions and then injected with ten percent formaldehyde. Eventually, they were encased in plaster to make accurate molds of both, and then were finally cast in latex rubber; hand-painted and added to the new reptile gallery. Terry personally cast and paintings of many of the reptile models for the gallery. Most of these reptile models are still on display in ROM galleries.

While carrying out the production of the reptile gallery, Terry told me of a recent collecting trip he and Dr. Scott of the ichthyology and herpetology department had taken to the Trinidad jungle. There they collected a living, full-grown, and extremely poisonous bushmaster snake. This reptile, too, was eventually molded, cast in latex, and hand-painted to become the main subject in a small South American diorama in the reptile gallery. Rather unique additions to this display were replicas of leaf cutter ants moving through the dense South American forest. This was accomplished by hiding a small electric motor connected to a bicycle chain; decorated with an assortment of reproduced Trinidadian leaves. When stepping on a switch hidden in the floor directly in front of the viewing window of this small diorama, the leaves attached to the bicycle chain rotated creating a convincing impression of leaf cutter ants moving through the jungle.

Later on, I spent a lot of my spare time with Terry. Much of this time was when he and his assistant were preparing to travel to Africa. They were to collect and document African wildlife and related vegetation. All of which, once shipped to the ROM, were to be used to produce two African dioramas in the museum's mammalogy gallery.

Following his African trip and the completion of two aforementioned dioramas, Terry's next adventure was to travel to the Galapagos Islands. Again, Terry was to design and build a diorama representing these unique Pacific Islands. On this voyage, he was accompanied by North America's most famous bird watcher, Roger Tory Peterson. Peterson was, and still is, best known for his field guides of birds which he wrote and illustrated. Apparently, while Terry was giving Mr. Peterson a haircut in the Galapagos, a Darwin Finch collected some of the hair cuttings to line his nest. The finch certainly could not have chosen a more appropriate nesting material.

During the planning sessions for the Galapagos expedition, I learned a lot about the flora and fauna of these islands. On his return, Terry constructed a full-sized Galapagos diorama which included several Galapagos sea iguanas. These were cast in latex rubber, from the collected and preserved specimen. I was fortunate to assist Terry in preparing some of these reptiles. I did this at home. One Sunday afternoon, while taking a break from painting sea iguanas, I happened to look out of our front living room window and saw a lot of cars stopping in front of our house. I was curious and wanted to find out what was going on; so, I went outside and soon discovered that my son Mark, six or seven years old at the time, had tied a long string to the tail of one of my iguanas and had placed the rubber reptile on the middle of the road. Hiding in some front yard bushes, he slowly pulled the string making the artificial lizard move slowly towards him. Passersby became curious, stopped their cars to see what this strange animal was doing in the middle of the road. I found this prank hilarious but told Mark that it was best not to continue his prank because it might cause an accident.

After the Galapagos expedition and the construction of the diorama, Terry's final diorama collecting trip was to India. This trip resulted in the construction of a major diorama displaying Indian tea plantation including some related Indian wildlife, such as several junglefowls.

I learned a lot about the work Terry had accomplished in his earlier years at the ROM. He illustrated several major books and published numerous scientific papers; in the latter, he described several new species. In the book *Birds of Ontario* by Dr. L. Snyder, head curator of the old ornithology department, Terry used both scratchboard and pen and ink drawings to illustrate each Ontario species. Later, Terry wrote and published several more books on birds. The book: *Not as the Crow flies* was illustrated and written by Terry and published some years later. It is truly a delightful read and is about some of his many ROM field adventures.

During the first year at the ROM, I got to know many other ROM staff members. One of these was Stu Downie, who had, for some time, worked in the Department of Mammalogy. Stu only had one leg and walked with a crutch under the arm on the side of his missing leg. Stu was able to keep up with anybody and had participated in numerous ROM collecting expeditions over the years. Stu was a cigarette smoker, and once in while I would ask him for one. Reluctantly, he would give me one and when I asked him for a light, he always said, "Christ, you want me to smoke it for you, too?" Unfortunately, one day Stu was gone from the museum. Apparently, he had been caught drinking alcohol on the job and was fired immediately. I was saddened and missed his presence.

Before I knew it, it was late spring and Dr. Crossman asked me to accompany him on a field trip. This was going to be my very first collecting trip for the ROM. At the time, Dr. Crossman (affectionately known as E. J.) worked with the family of fish '*Esocidae*,' pike-like fish and at the time was especially interested in *Esox vermiculatus* – the grass pickerel. He took me down eastern townships of Ontario, near the town of Brockville, where we collected in Jones Creek and some of its connected waters. We would leave early in the morning and return home after sunset. On the way east to Jones Creek, we passed by numerous large fields inhabiting herds of both dairy and beef cattle. In many of these fields, large full-grown dead elm trees cast long shadows in the early morning light. It was rather sad to see these once magnificent trees, killed by the Dutch elm disease. Soon there would be none.

Arriving at our desired location, we saw that Jones Creek was lined with knee-high grasses and a variety of shrubs. The creek flowed slowly, winding its way through clusters of trees and some open grass-covered vistas. We put on hip waders and dragged a thirty-foot seine through the creek and later through a large pond connected to it. The pond was some distance down a dirt road a few miles from our original location. E. J. was quite a big man at six foot five and when we dragged our thirty-foot seine through this body of water, I often had a hard time keeping up with him. However, being the junior member of our two-men collecting team I usually got to drag the net through the deep end of any given waters and in doing so frequently managed, without any effort, to fill my hip waders with chilly water.

By noon on my first ROM collecting trip, we had collected numerous species of fish from Jones Creek and its connecting pond. These included numerous mature grass pickerel and several species of minnows. Some of the latter were *Chrosomus eos*, the red-bellied dace. At that time, I had never seen living examples of these brilliant red-colored fish. These reminded me more of some exotic aquarium fish and not of a common Ontario minnow.

I always enjoyed these trips with E. J. to Jones Creek and its connected waters; over the next couple of years, we made several of these excursions. At times, instead of returning home on the same day, we made an overnight stop at the Kingston University field station. The main lodge of the station was located on a small lake which was filled with several species of colorful sunfish and some smallmouth bass. One morning before setting out to travel back to the ROM, I decided to do a bit of fishing. I didn't have any hand-fishing gear; so, I took a bit of sturdy string, tied on a hook made out of a safety pin and baited it with an earthworm. Surprisingly in a short time, I caught two small-mouthed bass. These made a terrific breakfast for the two of us before our trip back to the ROM.

I should add that on many trips to Jones Creek, we carried a packed lunch with us. This was prepared by my dad – a master chef. Sometimes, he packed each of us a half deboned roasted chicken wrapped around one of the chicken's drumstick, making a sizable chicken snack of both dark and white meat, accompanied by some homemade potato salad. This combination always provided us with a delightful lunch.

I guess both E. J. and I had strong personalities and at times we had our differences. However, in spite of this, or perhaps because of this, we became good friends. Often early on some Saturday mornings, E. J. would pick me up, and accompanied by his beagle dog, we would take off to hunt rabbits. We usually trekked through woods and fields in the Stouffville/Markham of southern Ontario. Enjoying each other's company and the fresh cool fall morning air of the great outdoors. We never had all that much success collecting rabbits but I think it was really more about getting out in the fresh country air and observing the sparsely snow-covered fields.

Over the years, E. J. and I made many collecting trips. One of our major trips happened in the summer of 1964. We drove down the eastern seaboard of the USA all the way to Alabama; eventually driving back up to Toronto through the Mississippi valley. On this trip, before entering the USA we stopped at E. J.'s uncle's farm in southwestern Ontario. His uncle had been ill and E. J. wanted to pay his respects. Apparently, E. J. worked on this farm picking tobacco during many of his summer vacations during his student days.

Following the short visit, we continued southwards and by the time we reached North Carolina, our 1958 museum Chevrolet station wagon broke down. We lost a day having the car problem resolved in Raleigh. The next day, we continued south and ended up staying for a few days with a friend of E. J.'s in North Carolina. This person was the district biologist for the State and provided us with an enjoyable and rewarding stay. For whatever reason, our host always carried a fully loaded 45-caliber handgun under the front seat of his car – somewhat unusual to say the least. Anyhow, the biologist took us on several excursions showing us collecting sites in North and South Carolina where we later successfully collected some of our thoughts after specimens of grass pickerel. The North Carolina countryside was quite different from what I was used to. We pretty much stayed in the coastal plain, the largest area of the state. Here, the sandy soil provides an ideal growing environment for stands of pine trees whose large cones littered the grounds. We did not visit any of the higher elevations in North Carolina, such as the Appalachian or the Smoky Mountains. These were not part of our designated collecting areas.

With a thirty-foot seine or with an electric shocker strapped to our backs we managed to collect a good number of grass pickerel. The electric shocker had, at one time, a bit of a problem. While strapped to my back and with the electrode engaged penetrating the water through which I was walking, I suddenly got an electric shock through my shoulder blades. Not a comfortable experience. When I told E. J. about the problem, I don't think he really believed me. He too tried the shocker and experienced the same results. After receiving a similar electric shock, E. J.'s response was, "Oh for heaven's sake." Anyhow, we resolved the shocker problem and soon managed to collect some grass pickerel.

The electric shocker did not kill any fish, it only stunned them making them float belly up to the surface of the water. We could then easily pick up all desired species with a dip net and then preserved these in ten percent formaldehyde.

I really enjoyed our stay with E. J.'s friend, the district biologist. I learned much about the flora and fauna of these parts of the USA. Each day I got to better understand the southern drawl, often fondly trying to imitate it. I soon got accustomed to 'grids' which was served with most meals. The latter I had not tasted or heard of before. It was a totally new experience for me.

From North Carolina, we continued southwards. We stopped in a couple of places in Georgia but had little success in collecting any of our desired species.

By now, the weather had turned very hot and super humid. It was the middle of summer and temperatures were consistently in the mid-nineties Fahrenheit with unbearable humidity – at least for a couple of northerners.

By the time we reached Alabama, the temperatures with high humidity remained pretty much constant in the nineties. In Alabama, we required special permission from the head of the forest department to collect in local waters. To get collecting permission we had to find the office of the man in charge of collecting permits. This proved to be a bit of a challenge. The head's office was located in a rather strange place. The entrance to his office was located directly behind the judge's chair in one of Birmingham's courtrooms. After finding this rather elusive man and after answering a variety of questions related to what we wanted to do, he gave us collecting permission and assigned two local environmental rangers to accompany us on our collecting excursions. Each of these rangers wore an open holster fully loaded with a forty-five caliber gun strapped to one of their legs. These rangers drove us around outside of the city limits and showed as several potential collection sites. At one of these sites, a black couple and their children were angling. I walked over to them with the intention of having a chat and seeing what they had caught. These people were somewhat reluctant to talk with me and our accompanying rangers did not seem all that impressed about me doing so. I found all of this a bit strange at the time, until later, when I realized that everything in Birmingham was still very much segregated. Sidewalks, shops, and restaurants were clearly labeled 'Whites only!' On one occasion, when E. J. decided to have a drink from the marble water fountain in a nearby park, I joined in and took a drink from a small copper pipe extending out of one side of this marble fountain. I was soon reminded by some passing white person that this copper pipe was to be used by Blacks only! What the hell – it certainly did not do me any harm and I made a point of drinking from that copper pipe every time I passed it.

The following day, after being introduced by the Alabama rangers to some potential collecting sites we began dragging our seine through some narrow four to five foot deep narrow creeks. The rangers had warned us that poisonous snakes were quite common in and near these waters. Apparently, cottonmouth snakes were quite common in our collecting sites. This did not exactly encourage me to jump into the creeks to drag our seine through these unfamiliar waters. We wanted to catch some more grass pickerel to add to the distributed records of this fish species. When carrying the electric shocker on my back, I felt somewhat braver knowing that I could zap any snakes I might encounter. Fortunately, we never did see any of the poisonous reptiles and did manage to catch several mature grass pickerel. These narrow and relatively deep creeks were, in places, lined by mature trees but for the most part flowed through knee-high grassy meadows without any shade. Collecting in the sweltering Alabama heat and humidity, we drank lots of water and bottles of ginger ale without ever having to relieve ourselves. By the end of our first day, we had managed to collect numerous mature grass pickerel and E. J. was delighted and satisfied with our day's work.

In total, we spent the best part of two days collecting in hot Alabama, after which E. J. decided that we had accomplished our goals and we soon got ready to drive back north. By this time, I really had become quite used to southern drawl and enjoyed hearing it. Before leaving Birmingham, and I forget for what reason, we stopped in a local hardware store to purchase whatever we needed for our return trip. While in the shop, a young boy, perhaps ten years of age, entered the store and was asked by the store's proprietor in a rather loud and joyful voice, "Did you all carry your daddy with you?" Expressions such as these I would not soon forget. E. J. and I would periodically try to repeat it with some sort of southern drawl to each other all the way back to Canada.

When it was time to get out of Birmingham, we were forced to make several detours. Some civil rights demonstrations and civil rights marches were interrupting and delaying traffic in all directions. Eventually, we did manage to leave town and drove north on one of the interstate highways. After several days of driving north, we deposited our southern collections at the ROM back in Toronto. I think both E. J. and I thoroughly enjoyed all of our experiences of the trip and were satisfied with our accomplishments.

Back at the ROM, things soon got back to normal. Building storage boxes and sorting and identifying fish, and keeping up with all that had to done on a daily basis. Dr. Scott was working on the completion of a book: *The Atlantic Fish of Canada*. I added a large aquarium at the northern end of the hallway facing the entrance of the department. In this, we displayed some live grass pickerel and from time to time various other native Ontario fish species, creating an interesting and inviting welcome to the department.

The following spring when the 'smelts' were running in Lake Ontario, I took a thirty-foot seine and made my way to the lake in the eastern townships with a friend and just before sunset we made several swoops with our net and caught buckets of these small silvery fish. We dropped off a good portion of these to E. J., W. B. and a few other friends who enjoyed eating fresh smelts. Also, from time to time, after hours, E. J. and I would go and collect some live minnows. These we usually caught in Duffins Creek at Brook Road and just north of highway two, in Pickering. These were used to feed any live specimens kept in our holding tanks, in the live room and to those fish kept in the large tank at the entrance to the department.

One day, on one of those humid, hot days in mid-summer, when you could literally see the alcohol evaporating from any opened bottle of preserved specimens we received a large load of several species of fresh whitefish from Lake Erie. Piled high on the laboratory table, all of them looked pretty much the same to me. They were all about the same shape and size and were shiny, silvery in color. Because of the sweltering heat of the day, we needed to preserve these specimens as soon as possible. In no time at all, Dr. Scott visually identified all of the whitefish and I got to formally label and preserve them immediately. Not too many ichthyologists could have done what W. B. did that day. Identifying such a variety of species of one family of fish, without the assistance of any artificial fish identification keys or reference books is quite impressive. Once more I learned a lot about Ontario fish species by listening to W. B.'s verbal comparisons and description of these whitefish species. Such significant donations to the ROM fish collection happened frequently and besides the daily goings on, kept the fish laboratory hopping.

That same summer when one day I was walking through the basement gallery by the student entrance of the ROM, I passed a familiar looking fellow. I called out to him, "Jerry," he stopped and tried to identify me. He hesitated and when I said, "Jerry McIntosh," he still seemed to be puzzled. I then told him, "Remember me, the German kid in your class with Mr. Hunt at Maurice Cody school." He then, although vaguely, remembered me. We shook hands and reacquainted. What a surprise to meet someone from a long time ago. Jerry had been hired by the ROM as one of the gallery designers and stayed at the museum for many years accomplishing some fine gallery designs.

That summer we had more than our usual share of hot, humid days. We survived the weather and before I knew it, all had passed. In early January of the following year of 1965, I was offered, by Dr. Scott, to participate in an ocean cruise. I was to collect fish in the Gulf of St. Lawrence off the east coast of Canada. However, lucky for me, this cruise ended up sailing south, down the Gulf Stream to the Bahamas, where we collected all sorts of Atlantic Ocean fish.

It was on a typical eastern Canada January morning, snowy and cold, when I set off by train to Montreal and from there caught another train to St. Andrews, New Brunswick. I had to wait for several hours in Montreal to catch my connecting train destined to McAdams, New Brunswick. After a cold, noisy, and rather shaky ride with several stops at small towns on the way, I ended up in McAdams. From there I took a bus to St. Andrews and made my way to the Biological Station. St. Andrews too was covered in snow. At the biological station, I met Dr. Jim Becket who was in charge of our upcoming cruise. Jim was not much older than me, and we hit it off quickly. Our research vessel was to arrive from St. John's, Newfoundland the following morning. This gave us time to gather up all of the necessary gear for the southern cruise. I stayed overnight at the Shiretown Inn. Since it was January, I was the only guest at the hotel. That evening, St. Andrews' mayor and the town council had a meeting in one of the dining rooms, immediately next to where I was eating my dinner. Before starting their session the national anthem was played on an old-fashioned record turntable with all council members standing at attention. Before turning in for the night, Jim Becket and I sipped a couple of glasses of home-brewed beer at his house and went to the St. Andrews estuary in Passamaquoddy Bay to hunt for ducks. This was apparently a favorite pastime during the winter month for many residents in St. Andrews.

The following morning the research vessel arrived. It was a large black painted dragger complete with a large otter trawl net stored on its after deck. Soon, everyone who participated in this cruise helped to load all of the required gear on board and by late afternoon we were on our way. Early next morning we arrived in Halifax, Nova Scotia where we stopped and picked up several hundred pounds of frozen mackerel. These were to be used as bait on our loglines which would be set once we arrived in the collecting areas of the southern Atlantic.

The A. T. Cameron continued southwards traveling the Gulf Stream towards the Bahamas and to Miami, Florida. Our large black stern trawler made good time until we hit the seas off Cape Hatteras. Just south of the Cape we ran into a major storm. The Cameron's bow pointed into the wind; our trawler bobbed up and down and vigorously rolled from side to side. Substantial waves crashed over the ship's decks; at times hitting the pilot house a story above the main deck. One of the challenges we faced was taking the ocean's temperatures every four hours to determine when we had reached our desired collecting locations. We accomplished this by tying a rope around our waist and securing this rope firmly to the rails of the ship.

The purpose of our cruise was to collect swordfish larvae – which survived in sargassum weed floating in the southern Gulf Stream. To collect these larvae, we trawled plankton nets off either side of the ship's decks; skimming about one foot through the surface ocean waters. These nets gathered a bunch of the sargassum weed in which numerous immature fish species made their homes. Occasionally we collected a swordfish larvae and each time we did so, we quietly celebrated this catch. All small, immature specimens were immediately preserved in mild solutions of formaldehyde and stored in 24-ounce bottles.

It was also our aim to collect mature swordfish to test these for mercury. The presence of mercury was of much concern in those days to the scientific community. To catch large swordfish we set long lines. These steel lines were several hundred feet in length with bright orange-colored floating buoy connected to each end. Long lines were usually set in the late afternoon and hauled back aboard the following morning. From each long line, at intervals of about thirty or more feet apart and twenty feet or so in length, steel-hooked, mackerel-baited lines were suspended to attract large ocean fish. All specimens caught on these lines were hauled over the ship's railings, unhooked, and dragged to various areas of the deck, to be processed and made ready to be shipped to the ROM from St. Andrews at a later date.

Our long lines collected several mature swordfish and an assortment of other large species. We caught a lot of sharks, including a couple of species of hammerheads, makos, black tips, white tips and several tiger sharks. We also caught a couple of bluefin tuna, a large sailfin, several dolphins, and a couple of good-sized marlins. Each time the long lines were retrieved and all large specimens were unhooked, the decks of the A. T. Cameron became slippery and congested. The mucus sloughing off the collected specimens made walking on deck rather hazardous. 'Proceed with caution' was the name of the game. I cut opened the bellies of all collected sharks and examined and recorded their stomach contents. I also removed all jaws from all sharks, and sun-dried these to prepare them for the ROM collection. Cutting out shark jaws proved at times to be difficult. I frequently had to re-sharpen my knives to accomplish this. The sharp denticles on the sharkskins frequently dulled my cutting tools. Some of the collected sharks I injected with formaldehyde, wrapped them in layers of formaldehyde soaked cheesecloth and stored them in sealed plastic tubing. All large specimens were measured and all examination results were recorded. At least one specimen of the various other large fish collected were also injected, wrapped in formaldehyde soaked cheesecloth and stored in sealed plastic tubing ready to be shipped to Toronto once back in St. Andrews.

We also twice use the large otter trawl with which the Cameron was equipped. Our first large trawl was in the Gulf Stream before reaching the waters of the Bahamas. We dragged the large stern trawl net over the ocean floor and caught a diverse variety of medium-sized fish. I preserved all of these to be shipped back to the ROM for the ichthyology collection. In our first trawl on the way to the Bahamas we unintentionally caught quite a few lobsters. A couple of these crustaceans were almost surreally large. One of them carried a 30-inch-long dogfish shark in one of his claws advertising his size and strength. Needless to say, that night on board we had quite a feast. However, the large lobsters proved to be rather tough and eating them was a bit like chewing on a tennis ball.

We continued long lining and all the retrieved large specimens once released from their hooks were organized on deck and I spent the rest of the day processing the catch. I measured all specimens, recorded their girth and their total length etc. Most of the swords were removed from large swordfish and the odd large swordfish I injected with ten percent formaldehyde to be eventually become part of the ROM collection. The same was done with all of the other large billfish. I removed all shark jaws from all collected sharks, cut open their bellies, and recorded their stomach contents. Removing the shark jaws again proved to be not all that easy and I continually had to re-sharpen my cutting knives. All collected specimens were labeled, their statistics were recorded, and they were prepped to be shipped to the ROM.

Working on deck under the burning sun was not always all that comfortable. Wearing nothing but a pair of shorts and sweating buckets soon turned my skin as brown as a milk chocolate bar. Sometimes just before sunset, when all of the work was done, I would catch a few flying fish with a regular fishing rod. You were always sure of their presence when you could see them breaking out of the surface of ocean swells and then gliding some distance over the surface waters. When successful, I immediately filleted my catch, sprinkled them with salt and pepper and fried them in melted butter for a tasty late-night snack.

By the end of January, we had completed our southern voyage and sailed back north to St. Andrews, New Brunswick. After reaching St. Andrews, and the Biological Station, I packed boxes filled with many glass bottles each filled with a variety of small fish ready to be shipped to the Royal Ontario Museum.

Back at the museum, all of the specimens collected on the A. T. Cameron arrived in our department. I unpacked, accessioned, sorted, officially identified, cataloged, and added the Cameron collection to the designated areas in our collection room. I can't exactly remember how many swordfish larvae we collected but I think it to be at least ten. Not bad, considering that at any given time the plankton nets skimmed only four-by-one-foot area of the Atlantic Ocean's surface.

My day-to-day departmental work soon continued. Fish were sorted; identified, and if needed, added to the collections. We received numerous loans of specimens and sent out just as many to other institutions. I continued to build storage boxes and filled many of these with larger specimens, many of which I had collected on the last southern Atlantic cruise. Before you knew it, the northern wall of the ichthyology department's hallway was lined with the storage boxes extending past the general zoological laboratory, as far as the entrance to the department.

In the second smaller collection room, many five-gallon bottles filled with mostly Ontario fish still needed to be transferred from formaldehyde to sixty-five percent alcohol. These fish also needed to be sorted and identified and many of these were added to the fish collection. Doing this work, I discovered that I had a real talent for breaking five-gallon bottles, especially those filled with formaldehyde. Every time I broke one of these large bottles I created an uncomfortable mess. The fumes of the formaldehyde soon permeated the air and I was forced to wear a water soaked cloth tied over my mouth and nose during the cleanups of my misadventures.

There were also many bottles of all sizes, on shelves and in cupboards throughout the department. These had to be sorted and their contents were to be processed. The fish contained in these bottles were mainly assortments of Ontario fish. Many of these specimens, too, were still stored in formaldehyde and needed to be washed in running water and transferred to sixty-five percent alcohol. A lot of these fish were added to the main fish collection. While spending time sorting and identifying all of this backlog of unidentified specimens, I really got to know the local Ontario fish fauna. Soon I could identify a lot of Ontario species by sight without using artificial keys.

Much of the department was soon becoming well organized. One day Dr. Scott called me into his office and in his usual enthusiastic voice asked me, "Pete, how would you like to go on another cruise?" This one was to collect fish in the Gulf Stream and in the southern Atlantic. I accepted immediately! This cruise, too, was designed to collect mature and larval swordfish and anything else we were able to get a hold of. We were to make a stopover in Puerto Rico and in Barbados. Once we reached Barbados I was to get off the ship and fly back home. In Barbados, Dr. Scott would join the cruise and collect fish on the way back to St. Andrews.

This time I was allowed to fly to St. John, New Brunswick. From there I was to take a bus to St. Andrews and on to the Biological Station. Again, Jim Beckett was to be the chief ichthyologist on board and again we worked well together and accomplished our goals. The ship on which we sailed, however, was very different from the Black Trawler I had taken on my first ocean voyage. This research vessel named the 'Sackville' was a converted World War 2 minesweeper which proved to be a very lively ship once out on the ocean. She was a rather narrow ship and dove and rocked vigorously out on the ocean waters.

Again, we loaded all of the required equipment and gear on board the day before setting sail, for southern waters, and once again we stopped in Halifax, Nova Scotia to load up with several hundred pounds of frozen mackerel. This time, we did not run into any storms and sailed south without any interruptions. We soon started collecting in the southern Gulf Stream. We took surface ocean temperatures on the way down and once these became warm enough we began to trawl our surface plankton nets collecting great bunches of floating sargassum weed in which numerous species of larval fish were entangled. A few of these proved to be swordfish larvae. This time we also caught some lantern fish in our plankton trawls. These were quite a surprise. Lantern fish usually live in the depths of the ocean; however, a few migrate to shallower waters during nighttime hours. The lantern fish we caught had tangled themselves in the sargassum weed and surprisingly many were still alive when I removed them from the nets. All of them proved to be species of the *Myctophid* family. I immediately put the living lantern fish into a small tank filled with seawater to see if any of their photophores would light up. I guess by this time they were in shock and I did not detect any operational photophores.

Between our daily sets of plankton nets, we also set a couple of long lines. Their hooks baited with whole mackerels were released over the rails of the moving ship. These long lines with their large orange buoy attached to it at either end drifted overnight in the ocean and were recovered the following morning. Some large swordfish, several species of sharks, a couple of marlins, and a large sailfin were captured. We also managed to collect several large dolphins. These beautiful multi-colored fish were absolutely striking and I could not resist making some color sketches documenting their living colors.

While unhooking our catch from the long line, all of a sudden there was commotion near the starboard side of the rail. I ran over to the rail to see what this was all about. One of the deck hands had gone overboard and was treading water among the retrieving long lines hooked sharks and some other large fish. A lifebuoy with an attached rope was thrown overboard and the crewmember was soon retrieved and brought back on deck. Fortunately, no harm was done.

Knowing that I had completed the construction of two large eight-foot long storage boxes back at the ROM, I was able to process most of the large fish we collected by injecting them with ten percent formaldehyde and wrapped each in formaldehyde soaked cheesecloth making them ready to be shipped to the ROM at a later date.

While we were collecting and processing specimens, some of the Sackville crew was busy painting some of the ship's deck areas, including the ship's masts. One of the crewmembers was not thinking all that well on that particular day. While painting one of the ship's mast from bottom to top, he soon found himself trapped and had to slide down the freshly painted mast covering most of his clothes in white paint. Some days you just can't win!

We continued to collect using both methods: plankton nets and long lines off several Caribbean islands, and filled many boxes of bottles with immature fish along a fair number of lantern fish. I once again carved out several shark jaws, sun dried these, and recorded the stomach contents of all collected sharks. In the stomachs of a couple of tiger sharks, I found big chunks of melon rinds which some crewmembers had tossed overboard the night before. Probably not a good place to swim.

Continuing southwards on a bright hot sunny day, we entered the old port of St. Juan, Puerto Rico. The seas were quite choppy and our converted minesweeper bounced up and down and rolled vigorously from side to side which made entering the harbor a bit uncomfortable. Just before entering the harbor of Puerto Rico, I saw the old fortress built years ago to protect the island from unwanted invaders. After an hour or so of traveling through choppy waters, we entered the harbor of an American Naval base and parked the Sackville directly across from an American naval destroyer. This made our ship look rather inadequate and puny. The sailors on the American destroyer were all decked out in their whites, celebrating some sort of American holiday.

Once the Sackville was secured to the dock, we went to visit the town of St. Juan. We toured its old streets and enjoyed some of the local cuisine along with the historical sites of the island. We also visited the Bacardi Rum distillery, just outside of St. Juan. There we could not resist purchasing various shades of rum for the sum of $2.40 a quart. It had to be done! We each stocked up on several quarts of Bacardi knowing full well that it was impossible to drink all of these purchases before returning home. We had to devise some sort of system to get our acquisitions of rum into Canada without paying duty. An idea for doing so came to me quite easily. I decided to fill some 32-ounce bottles with rum and in each of these, I inserted a small sealed plastic bag containing a small preserved fish. This proved to be very successful. Canadian customs looked at these rum filled bottles and accepted these as just more bottles of preserved fish.

We spent three days in Puerto Rico and after refueling, we continued sailing south, collecting fish off several Caribbean Islands, finally docking in the harbor of Bridgetown in Barbados. Once in Bridgetown, I visited the local market which was located not far from where we had docked. As a bit of a 'foodie,' I thoroughly enjoyed inspecting all of the fresh produce, especially all of the variety of locally caught fish.

I'm not sure who, but someone told us to visit a certain nightclub in Bridgetown. Apparently, it was unique and had to be seen. The name of the establishment was Harry's Place. The nightclub was located on the second floor of an old house. On entering Harry's Place we soon discovered, to our amazement, that all the waitresses served their customers topless. Unfortunately, we had to depart Harry's Place in quite a hurry. This was due to one of our companions from the Sackville who had a bit too much to drink and decided to make a bet with the proprietor of the establishment, presumably Harry, which was if he took all of his clothes off and danced with the girls naked on stage, he could have his pick of any of the girls he danced with for the rest of the night. Well, guess what, he did exactly what was asked of him and before we knew it, our companion was up on stage with a number of girls wearing only a pair of socks. Quite a sight! When the time came for our drunk friend to have his pick of the girls, however, the proprietor had changed his mind and decided not to honor the original agreement. Things soon got ugly. Our friend insisted on the agreement and some pushing and shoving ensued. A group of large men escorted us hastily, two stairs at a time, down to the street.

The following morning, I met Dr. Scott at the Flying Fish club in Bridgetown, where he had been staying. We spent that day and evening at the club discussing my voyage and what we managed to collect. I informed Dr. Scott that the Sackville was not the most comfortable ship to sail on. I don't think he quite believed my explanations of my Sackville experiences. However, I found out later that W. B. too did not like sailing on the Sackville. Instead of sailing back to St. Andrews as originally planned, he got off the ship in Bermuda.

After spending a couple of days in Barbados, I flew back to Toronto and soon got back to my normal routines at the ROM. I built more storage boxes and continued to sort, identify, and catalogue specimens retrieved from various storage places within the department.

The Sackville collection arrived at the museum in good order. I accessioned, sorted and officially identified all of this collection and washed the formaldehyde-preserved specimens in running water before transferring them into sixty-five percent alcohol. We did manage to collect a dozen or more swordfish larvae, and quite a few other larval fish. The collected lantern fish and a variety of large, mature specimens made some unique additions to the ROM fish collection.

Back in cold Toronto, I realized how nice it was to cruise the southern Atlantic in the middle of a Canadian winter. It made the winter pass more quickly and before you knew it, winter was over.

By the beginning of that spring, I had been in the Department of Ichthyology and Herpetology for some ten years. How times flies when you are having fun! The Vietnam War was still going strong and pretty much filled all of the daily news reports both on radio and television.

This particular spring, Dr. Scott wanted me to accompany him to collect a specific Ontario minnow species that being *Notropis umbratilis*, the minnow. This particular species of Ontario minnow inhabited small streams in the Lake Erie region of Ontario. During spawning, each spring, the males of *umbratilis* turn a gorgeous bright blue. We left for our designated locations near Lake Erie early one morning. The weather was still cool but sunny. We had loaded a thirty-foot seine, hip waders, some ten percent formaldehyde, and a few twenty-four-ounce bottles into W. B.'s car and set off for Port Dover on Lake Erie. On the way down to the shallowest lake of the great lakes, I again noticed the many large dead elm trees standing bare and sullen in grass-covered fields. Again, all had been killed by Dutch elm disease, an invasive, highly infectious fungal disease which started spread throughout southern Ontario in 1946. On the sides of the highway, amongst the wild grasses, clusters of wild phlox flowers, some white and others in various shades of purple, were in full bloom. All waving in a gentle breeze among tall grasses and spent dandelions. A delightful morning to be away from the office and laboratory for a while.

Arriving in Port Dover, W. B. wanted to visit an old friend – the owner of a commercial fishery. He had worked with this man years before when he was examining and documenting the Lake Erie whitefish population. At the fishery, all sorts of freshly caught Lake Erie fish were on display and we could not resist purchasing several pounds of filleted Lake Erie perch. Port Dover was, in the past, best known for its Lake Erie fisheries, of which only a few remain.

By early afternoon we were seining narrow streams, often hidden from the road by tall grasses and a multitude of other low growing vegetation flourishing vigorously in the spring atmosphere. From a couple of these streams, we managed to collect several *Notropis umbratilis* of both sexes. The males in full color – brilliant blue. With the *umbratilis* we also caught several other species of minnows which would be identified back at the ROM and most likely added to the ROM fish collection. As always, I got to drag our seine through the deep end of the streams. And as usual, I managed to step into some deep holes filling my hip waders with cold water. W. B., being a very practical person, brought with him a change of clothes. I, of course, did not. Always being a total gentleman, W. B. offered me his spare dry clothes. Standing along the side of the road taking off my drenched apparel, I accepted W. B.'s offer and soon discovered that his shirt and pants did not fit me that well. They were somewhat small and I looked a bit odd with the pant legs not reaching past the upper part of my socks. Good old W. B. got quite a kick out my appearance and soon demonstrated this by insisting that on the way home, we stop for a cup of coffee at the nearest restaurant.

The coffee shop customers too enjoyed my appearance and looked me over with bewilderment. I truly enjoyed that collecting day with W. B. and again learned a lot from the ichthyological expert.

Later that year Dr. Scott and his wife started to make trips to Barbados to collect Caribbean reef fish. In fact, W. B. collected many of these and soon quite a backlog of various sized bottles filled with a diverse variety of unfamiliar species had been deposited in the ichthyology department. The tropical reef collections grew in leaps and bounds, adding a great variety of new species to the collection.

Through the generosity of W. B., I was allowed to make a bit of extra money by working Saturday mornings, sorting and identifying many of these reef fish collections. By now I had been involved in a couple of ocean cruises, had collected fish in various regions of Canada and the USA; my knowledge and understanding of the world of fish had certainly grown immensely and my life in the ichthyology and herpetology department continued in full swing. Each year, usually by the end of June, our summer students returned to clean and top up all of the bottles and containers stored throughout the department. Their antics, their wit, and their sense of humor continued to add much delight to our daily routine. Greasing door handles with Vaseline was one of their many pranks. At times, a pail half filled with cold water was balanced gingerly on a partly opened door. Anyone who entered through this door would be soaked and surprised. During this particular summer, a new office secretary had been hired who, to say the least, was a bit gullible. Of course, our summer students caught onto this quickly and immediately engaged her in one of their pranks. Early one morning, the boys decided to test her gullibility. They took a small preserved frog out of the herpetological collection and put it into a one-gallon bottle which they covered with a white cloth totally hiding both the jar and its content. Our students convinced the new secretary that this little frog in the bottle hidden under the cloth would grow at a tremendous rate throughout the day. Each time the secretary stepped out of her office the boys quickly switched the preserved frog to one a bit larger preserved frog and each time after the switch showed the secretary how much the frog had grown. By the end of the day, the boys had managed to squeeze a huge, full-grown Canadian bullfrog into the one-gallon. This specimen was at least ten times larger than the original preserved specimen. I suspect that it took some time for the young lady to discover that all of this had been a joke.

Each summer, while the students were working in the department to break the monotony of cleaning and topping up bottles, we collected a few dollars from each staff member and just before lunchtime made our way to the Kensington Market. There we bought a couple of loaves of freshly baked bread, a variety of cheeses, some pickles, and a couple of pounds of sliced cold cuts. Back at the museum, we served all of our goodies in the ichthyology classroom for lunch to all staff members. It was always a nice break from the usual, and it gave everyone in the department a chance to spend some time to compare their latest achievements.

By this time, E. J. had written and published a book: *Fish of Quetico Park*. The park, a wilderness area in

Northwestern Ontario, is well known for its many canoe trails and as an excellent fishing destination. When writing this book, E. J. asked me to do the illustrations of the fish native to the area. I drew all Quetico species in pencil, including a hint of each species' habitat. E. J. checked all drawings for accuracy and I completed all of them in a black and white halftone acrylic wash. These illustrations were right up my alley and added an important visual dimension to the publication.

All of this made for quite a busy and rewarding year. In July, I was ready for a couple of weeks of holidays. I told Dr. Scott that I was going out west for my holidays and he asked me, if he let me take a bit of extra time for my vacation, I would collect some fish for the ROM while I was out there. I agreed and I ended up collecting quite a few small Albertan fish outside of Calgary. Driving across Canada I always enjoyed the Canadian western landscape. Seeing the flora and fauna of the west, including mule deer and pronghorn antelope was always a delight. It awakened my everlasting interest in wildlife art. While I was out there, I drove down to the Canadian and Montana, USA border where I collected a fair number of fish species, all ending up in the ROM fish collection. On the way back to Toronto, I also managed to collect in some of the tributaries of the Red River in Saskatchewan. There too I caught a variety of small fish, mostly minnows. These proved to be valuable additions to the ROM fish collection.

Having completed another interesting and successful year, in early January of the following year, I was asked once again if I wanted to participate in another southern ocean cruise. This time, there would be two participants. In the second week of January, Dr. Crossman and I set off for St. Andrews, New Brunswick, to the Biological Station.

On this particular cruise, we were to collect in the southern Gulf Stream, and all the way down to Wilhelmstadt, Curaçao. Again, our main objective was to collect larval and mature swordfish and bring back anything else we managed to catch. Once again we loaded the ship with the required gear at the St. Andrews docks. This ship was one of Canada's premier research vessel, named the 'Dawson.' It was indeed a fine ship with all of the amenities one could possibly ask for.

On the way down, traveling the Gulf Stream we again intermittently recorded the surface temperatures of the ocean until these became warm enough to launch our plankton nets. We also set several long lines on the way to our destination. Both fishing methods proved to be successful and E. J. and I soon found ourselves working on deck, preparing all of our catches to be eventually shipped to the ROM from St. Andrews. One day while hauling back the previous night's long line, a couple of the Dawson's crewmembers had trouble lifting a large hooked swordfish over the ship's rail. When E. J. saw them struggling, he stepped up to them and said, "Excuse me, boys." He grabbed the swordfish by its bill and with one hefty swoop, landed the large fish on deck. All deckhands applauded vigorously. In such cases, E. J.'s size sure came in handy. However, I often felt sorry for him when he made his way around the inside of the ship. He frequently bumped his head on the many lower parts of the ship's ceilings.

We continued to successfully collect many different species, on the Dawson Cruise – both large and small. When we finally docked in Willemstad, I decided to take in the sights and sounds of this interesting European-style city. Many of the larger buildings, now somewhat neglected, must have been gorgeous in their hay days. I soon discovered that Willemstad was designated as a UNESCO World Heritage Site.

After a couple of days in Curaçao, E. J. and I spent most of one day packing up all collected specimens making them ready to be shipped to ROM once the Dawson returned to St. Andrews. Soon afterward, the two of us boarded a plane to Port of Spain, Trinidad. At customs in Trinidad, we experienced a short delay. E. J. was carrying a large sword which had been cut from one of our collected swordfish. The sword was tightly wrapped in cheesecloth and bound tightly with rope. The customs officer at the airport in Port of Spain insisted that E. J. unwrap the sword and to explain to him what this item was all about. Once all explanations had been accepted, we caught our next plane and flew to the Caribbean island of St. Lucia. E. J wanted to check out this location for future collecting sites. In St. Lucia we stayed at quite a posh hotel which E. J. labeled 'rip-off ranch.' Everything at this resort was expensive and we had a hard time finding some affordable food for ourselves.

After a couple of days in St. Lucia we flew home and a few weeks later, back at the ROM, all of our collected specimens from the Dawson Cruise arrived in good order. All specimens were accessioned, sorted, officially identified, cataloged, and added to the ROM fish collection. The Dawson Cruise was my last ocean adventure.

In the summer after the Dawson Cruise E. J. and his wife, Margaret, did return to the island of St. Lucia and collected a diverse variety of Caribbean reef fish, which added a lot more diversity to the ROM fish collection.

In the meantime a third curator, Dr. Allan Emery, was hired and became an active part of our department. He was given Dr. Dymond's old office to work from. He had recently completed his Ph.D. and in doing so had described a new species of reef damselfish. This fish he named in honor of Dr. W. B. Scott – '*Chromis Scotti.*' Allan was young, ambitions and brought a lot of enthusiasm to the department however, not all that long after his arrival he became a part of senior management, coordinating curatorial business.

It was not all that long after Allan's arrival, through his recommendations I was contacted by a Hollywood movie company who had decided to make a science fiction deep ocean adventure film in Toronto. Allan asked me if I was interested in assisting the movie people in training a variety of saltwater fish for the film. I was to make certain fish species, on command, to perform particular behaviors. I was, of course, very interested and Dr. Scott generously permitted me to leave the department for a couple of months and work on a movie set. I was to be paid twice the money I was making at the ROM and made a deal with the movie producers to inherit several large aquariums for the ROM once the movie was completed. I pretty much worked eight to nine hours each day, training various live saltwater fish to attack model submersibles and to swim in any desired directions. I accomplished these challenges by enticing any given species with some of their favorite food – mostly using frozen brine shrimp and/or raw meat. I was also allowed, from time to time, to use an underwater camera to film certain living fish in some of the large aquaria or in an above-the-ground swimming pool, both outfitted with artificially created underwater habitats. I must admit working on a movie set was most enjoyable and quite different to what I was used too. It was a nice break from the usual routine.

One of the stars in this science fiction movie was Ernest Borgnine, truly a very nice and very professional person. He would often clown around on set, telling jokes to whoever was around; however, when called on to perform for the camera, he instantly became serious and never missed a line.

My time with the movie company went by quickly, and within a few weeks I was back at the ROM continuing my real job.

Back at the ROM, both curators had become very interested in southern reef fish. They wanted to add diversity to the collection and to expand their knowledge of the world of fish. However, this was interrupted by the department receiving a large contract from the Canadian Federal Government to produce a major book on the freshwater fishes of Canada. All staff of the department soon became involved in this gigantic project. In fact, more staff was hired to accomplish this undertaking. I was promoted to senior technician. A second technician and a second secretary were hired immediately. Cheryl Goodchild was hired to work in the department's library and to assist the curators with all sort of related projects for the production of this major publication. Even the wives of both curators, especially Mrs. Scott, spent a lot of time assisting their husbands in gathering information for the book. I got to collect fish in Ontario where the ROM fish collection was sparse. One of those collecting trips took me to the Ottawa River where I collected in the Ottawa and several of its tributaries between Pembroke and Mattawa. On certain occasions, I would use rotenone a white liquid, extracted from rubber trees, which temporarily immobilizes fish when certain amounts of this liquid were added to any given waters. Fish residing in rotenone-treated waters flow belly up on the surface and then can be easily scooped up to be preserved.

On one occasion, when I had used rotenone in a small tributary of the Ottawa River, numerous fish of various sizes and species were floating belly up on the surface of this small stream. Soon a provincial police cruiser stopped abruptly on the shoulder of the highway where I was working. The officer stepped out of his cruiser and asked me what I was doing. I showed him my collecting permit and told him that I was investigating what was wrong with the fish in these waters. He accepted my explanation and soon drove off leaving me to my collecting. At the time I thought that my (not all that truthful explanation) was simpler and took a lot less time than explaining to him what I had done. I spent the best part of a week collecting in the Ottawa River area.

Once back at the ROM, my job for the book: *Freshwater Fish of Canada* started in earnest. I was to gather for the curators, all of the morphometrics for all Canadian fish species. For the next couple of years, I spent hours each day counting fish scales, fin rays and making proportional measurements from the smallest to the largest fish species found in Canadian fresh waters. Much of the above was accomplished by peering through a microscope. For certain species, I took x-rays and counted their vertebrates. Doing all of this detailed work I certainly got to know a lot about the freshwater fish of Canada.

Each species included in the book had to be scientifically and accurately illustrated. For this job, the curators hired Anker Odum, the staff artist of the ROM entomology department. At the time, Anker was illustrating a major book being written by Dr. Glen Wiggins, the head curator of that department. Anker was a superb pen and ink illustrator and ended up, on his own time, illustrating most species for the *Freshwater Fish of Canada* and he did indeed do a superb job. I too got to do some pen and ink illustrations for the project. However, I must admit that my pen and ink illustrations never turned out as good as Anker's. Pen and ink was never my forte. I was much better at full color and black and white halftone illustrations. Nevertheless, it was great fun working with Anker Odum. I learned a lot from him and at the same time really got to know Anker the man.

The work on the book *The Freshwater Fish of Canada* seemed to go on forever. I got to design the book's cover and was honored to be able to paint several full color illustrations of certain species which were added to the final publication. Eventually, this gigantic project too came to an end.

During the production of the freshwater fish book, I also managed to study design and compositions with some of the ROM's masters. I spent time working with both Terry Shortt and Anker Odum on producing a variety of wildlife art paintings.

By the time the freshwater book was completed, Dr. Scott spent more and more time in St. Andrews, New Brunswick. He built a house for himself and his wife, Milly, in St. Andrews and soon moved there permanently to become the head of the Huntsman Marine Laboratory. I became more and more involved with other ROM personnel. I would frequently meet with Terry Shortt and his assistant artist, Paul Geraghty, to discuss and participate in the production of some wildlife paintings. We often talked about wildlife art – past and present; criticized each other's works, discussed how to do things differently and how to improve our techniques.

During these times Paul Geraghty and I often traveled to the Queens University Field Station. These trips were led by Brock Fenton – a Ph.D. student in the ROM mammalogy department. In the area of the field station, in some small caves and in some attics of a few old houses we collected little and big brown bats for Brock's studies.

Meanwhile, I continued to spend much time with Terry Shortt, Anker Odum, and Paul Geraghty. All of us working on a variety of natural history paintings. I learned a lot from these seasoned veterans, and in 1968 together we produced a traveling exhibition of wildlife paintings. Each produced several full color works portraying animals related to our specific disciplines. I made half a dozen paintings of fish in their natural habitats. Terry painted some terrific bird paintings. Paul painted a number of mammal pictures and Anker created pictures of various insects, including some great paintings of butterflies. Our traveling exhibition was well received and for quite some time was displayed in various libraries, universities and other cultural institutions.

I learned much about drawing, painting, mediums, color, composition, and design by hanging out with these artists. It was at this time that I started thinking about some sort of future in the art and design field.

I became more and more involved in the production of wildlife art and came up with the idea to organize an international wildlife art exhibition at the ROM. Most conventional art galleries have never accepted works of the visual arts containing wildlife. These were, and still are, labeled 'scientific renderings' – no matter how skillfully and artistically these have been accomplished. When I suggested my idea to Terry Shortt, he immediately became very enthusiastic about such an exhibition and threw in his full support. At the time, I was unaware that Terry had been experiencing some difficulty in his life. Apparently, he suffered from serious depression and had lost total interest in working for the ROM. When I suggested the International Wildlife Art Exhibition to him, all these problems instantly disappeared and before I knew it Terry began laying out the groundwork for such a show. All I had to do was to convince ROM senior management to sponsor such an exhibition. Amazingly, we encountered little resistance. At that time Dr. Walter Tovell, the former head curator of the Department of Geology, was the director of the Royal Ontario Museum. He was very approachable and showed a keen interest in our proposal and soon approved the project. Dr. Tovell was a keen birdwatcher and long had been interested in the world of natural history. Dr. Tovell introduced Terry and me to a friend of his – a wildlife art collector. This man, David Lank, soon became totally associated with our wildlife art project and contributed much to the success of the exhibition. He wrote countless letters promoting and adding intellectual and interpretive depth to wherever this was required. All I needed now was permission from both curators of my department to work part-time on the project. Both, Dr. Scott and Dr. Crossman allowed me to do so. In fact, Dr. Crossman volunteered to take on the role of a curator in charge of the exhibition and assisted us in its development. At the same time, we received a lot and valuable assistance and support from the ROM's human resources department. Both Mrs. Downie, head of the department; and her assistant, Cathy McKay, constantly provided help whenever needed. Especially Cathy McKay, who constantly assisted us from the very start to the completion of the project.

Terry Shortt and I met every working day and sometimes on the weekends to plan the project. We soon became close friends and truly enjoyed plugging away at the Animals in Art exhibition. I worked, each day, on my continuous departmental obligations and spent some time on the art show project.

We set up a selection committee of Canadian and American contemporary wildlife artists, art gallery owners, and art collectors. Terry did most of the research for the exhibition going all the way back to the earliest European cave paintings. He also contacted many North American wildlife artists he personally knew. We managed to contact artists from all over the world. Some of these were quite well known in Europe but had not ever been seen or heard of in North America.

Through some European hunting magazines, which I received on a monthly basis, we got to know the art of two well-established European wildlife painters. Their work was often used to illustrate articles in these magazines on various hunting excursions. One of these artists was German and the other was Dutch. The first, Manfred Schatz, apparently a Stalingrad veteran, painted in oil paints mostly large canvases of European mammals and birds in action. Using broad strokes and well understood reflective colors he completed numerous large paintings the North American public was totally unaccustomed to. The other artist, from Holland, Rien Poortvliet, frequently demonstrated his personal experiences of the great outdoors in his works often revealing his sense of humor. We also managed to get a great representation of the granddaddy of all wildlife artists, Bruno Liljefors of Sweden. We were able to borrow several of his large canvases from the Swedish National Museum in Stockholm. These large paintings were truly a sight to be seen.

During our searches for artists' works, Terry Shortt made a surprising discovery at the Blacker Wood Library in Montreal. He discovered many life-sized Indian bird paintings rendered in the seventeen hundreds, by a female artist by the name of Lady Guillam. This artist was totally unknown and had, while her husband was stationed in India as part of the British army, produced numerous oil paintings of Indian birds. These were painted in full color in their local habitats accurately representing certain species. We managed to loan ten of her works to be displayed in our exhibition and as far as I know, her work had never been displayed to the public and has not been exhibited since.

Terry and I also traveled to Calgary, researching the wild art collection of the Glenbow Museum. I vividly remember Terry running down one of the hallways of the Glenbow, jumping up and clicking his heel in excitement. On our trip to the Glenbow Museum, we were accompanied by the ROM's head photographer Lee Warren, who took large-format pictures of all of the paintings we thought to have potential to be displayed in the ROM exhibition. All of Lee's superb photographs not only made a great reference to the Glenbow collection of wildlife art but were also used to illustrate the work of many artists in the exhibition's catalog.

During the planning of the show, Terry and I also visited New York City to meet several contemporary wildlife artists including Dr. Roger Tory Peterson. On one of the visits to New York City, we attended Tory Peterson's talk at the New York Natural History Museum. The introduction to his talk was rather unique. Three nude people, two males and one female, ran across the stage shortly before Peterson's lecture. This certainly got everyone's immediate attention.

It took Terry and me two plus years to plan and to get everything organized for the ROM Animals in Art exhibition. We managed to produce and display the most diverse and largest international wildlife art exhibition ever held – anywhere. The show turned out to be successful and very much appreciated by the general public. The many wildlife artists, from all over North America and England too got a lot out of the exhibition. The exhibition launched several wildlife painters who became totally independent, and from then began freelancing and made a living by painting wildlife art. This included Robert Bateman who after the ROM exhibition made art his full-time vocation.

On July 1, 1976, Terry Shortt retired from the Royal Ontario Museum. For me, it was a very sad day. A retirement party was held in his honor in the ROM garden, just outside of the old cafeteria. Dr. Peterson, head of the mammalogy department, gave the farewell speech to the many ROM staff and many outsiders who attended this gathering. Terry, however, did not attend his retirement farewell. He was somewhat disillusioned in which direction the institution was heading and from then on had nothing more to do with the institution. Things were indeed changing at the ROM and all of these changes were totally foreign to him. Terry never returned to the museum, not even for a visit.

With Terry retiring and Dr. Scott moving to the east coast the old place really did not seem to be quite the same. Terry continued to work with bird art and wrote in the following years on a variety of subjects several books on birds. Each of these he illustrated with some superb bird portraits each accomplished in transparent watercolors.

I think that Terry's final project at the ROM, the Animals in Art exhibition, revived his passion for and in bird art. We often talked via the telephone, discussing our past ROM experiences and accomplishments.

Following the Animals in Art exhibition, I think, as a sign of appreciation, Terry with the help of a well-known zoologist by the name of Dr. Stu MacDonald of the National Museum of Canada in Ottawa, set me up to travel to the eastern high Arctic. Without Terry's support and influence, I would have never gotten such an incredible opportunity.

It was about a year following the Animals in Art exhibition that I was preparing to travel to the far north. It was late June in 1976, when I bought the necessary clothing and gear for my northern Canadian journey. To make sure that I would return from the north our summer students plastered a large playboy calendar on the west wall of the laboratory, right next to my desk. This stood out like a sore thumb. This sort of decoration was a rare sight anywhere in the museum.

Once more, I needed permission from both ichthyology curators to take some time away from the department for my high arctic excursion. And once again, both curators were most generous in allowing me to participate in the high Arctic voyage. In late June that year, I flew to the Montreal airport where I met Stu MacDonald. From there, we flew in a jet to Resolute Bay and after several hours of flying north, we landed safely on a gravel runway at our first Arctic destination. Unfortunately, due to heavy fog we ended up stranded in Resolute for a couple of days. This gave me time to get used to the twenty-four-hour daylight and to acquaint myself with some of the northern terrain. There was still a lot of snow on the ground and the only sign of any wildlife were some snowbirds pecking in some exposed areas.

At Resolute Bay, in the cookhouse, I listened to many stories told by experienced Arctic travelers who were stranded with us before they too traveled to their designated areas of work for the short Arctic summer. Some of these stories told of some touchy experiences they had with polar bears. Not exactly encouraging, when I thought of traveling by myself and without any weapon at my destination of Bathurst Island.

When the fog finally lifted, Stu MacDonald and I boarded a Twin Otter plane which took us to the Bathurst

Island field station. Stu Macdonald had built this field station years ago, for himself and for others who wanted to do research in the far north. It was now early July, the weather was bright and sunny with little gusts of wind interrupting the otherwise total silence. I was soon reminded that I was in the high arctic. On the fourth of July, Bathurst Island was totally engulfed in a snow blizzard. Fortunately, this was only a one-day event.

Once I had settled in on the Island, I was pretty much on my own and was able to do whatever I wanted to. I lived in a cloth-covered pakole from which I traveled daily to various parts of the island. There were two students, working for Stu who lived on the Island for the summer months. We would only see each other in the cookhouse for meals at which time we compared our daily experiences.

I discovered how naive I was about all things Arctic. Having worked mostly with fish, I discovered my lack of knowledge of all of the wildlife living in the far north. However, once again this turned out to be a first-hand learning experience which would be difficult to experience in any other way.

I took long daily walks setting out from camp in all directions. I encountered many birds and their nests. I also saw numerous Arctic foxes searching for these nests up and down Bathurst Island hillsides. Once in a while, a fox found a bird nest and quickly consumed all within these nests; usually repelling the vigorous protests from their adult occupants. The snow was still melting; forming shallow ponds all over the wide-open tundra. Many of these melt ponds were occupied by a pair of snow geese or a pair of king eiders ducks. From time to time, I encountered small flocks of phalaropes that were nesting across the vast tundra. I rarely encountered any lemmings usually plentiful on Bathurst Island. Their population was really down that year and because of this, I saw no snowy owls whose survival pretty much depends on these small northern mammals. On one occasion, I briefly saw a weasel hurriedly searching for food. He soon disappeared in a nearby hollow. I saw numerous ptarmigans and they had totally changed into their summer plumage from their winter white. At times, I got quite close to these partridge-sized birds while they were pecking their way through patches of the stunted willows and purple saxifrage. Above, jaegers constantly flew swiftly; changing direction, searching for anything they could attack and feed on. Once in a while, I discovered a skeleton of muskoxen littering the open tundra.

After a few days on Bathurst Island, Stu invited me to participate in a couple of trips to some other Arctic destinations. On one of these excursions, we traveled by helicopter to a small island in the Penny Strait. Stu was looking for the breeding location of a specific species of gulls which was apparently secretive. The center of this small island was totally covered with several species of nesting seabirds. Whether or not Stu found what he was looking for remains a mystery. On the sandy shores of the Penny Strait Island, fresh polar bear tracks were common and we soon decided that one of us needed to carry a loaded rifle during our island investigations.

On another excursion, Stu took me all the way to Ellesmere Island. This time we traveled in a single Otter airplane, quite an old machine which had accumulated many flying miles throughout the eastern Arctic. On our way north, we landed on a couple of designated beaches to drop off a forty-five gallon drum of fuel to be used by other northern travelers. To accomplish the fuel drops, our pilot first tested the firmness of these sandy beaches. He ran the single Otter over the designated beach at full speed; confirming that it was solid enough to land on. If the beach was suitable to land on, the pilot turned the plane around and landed the single Otter. Once the plane stopped on the beach, we opened the side door and rolled out one of the forty-five gallon fuel drums. Each landing and takeoff seemed to be challenging for the pilot and for me was at times puzzling how the pilot could accomplish each of these landings.

About halfway up to Ellesmere, Stu pointed out an abandoned Gyrfalcon nest. Apparently, this nest had been occupied every year for many years until now.

After flying for several hours north, we landed on a flat strip of fine grass on Ellesmere Island. Soon after visiting and meeting the staff of the Ellesmere Island field station, Stu and I took a walk some distance from the station where we discovered a dead muskox. This substantial animal had not been dead all that long. Stu was puzzled why the muskoxen had died. There were no visible signs for its demise. Stu firmly grabbed the animal by one of its hind legs and with some effort rolled the musk ox over to see if we could find any visible injuries. We found nothing. While surveying the dead muskox, a white Arctic wolf made its way towards us and circled us several times at a distance of about ten feet before disappearing over the distant horizon.

We spent the best part of a day on Ellesmere Island. I admired its rolling hills covered in subtle shades of green.

Here and there some blooming Arctic flowers added splashes of white and shades of purple colors to the vast landscape.

Late in the day after refueling, we climbed back into the single Otter plane and started our long trip back to the Bathurst Island. We spotted a couple of polar bears searching over the still frozen sea for seals. We also saw a couple of small herds of caribou traveling north over the seemingly endless barren grounds.

Back on Bathurst Island, I spent another week walking over the ever-changing landscape. I had totally gotten used to the twenty-four-hour daylight and my Arctic beard was growing well. On my daily walks, when no wind was blowing, I realized that there was total silence and the only sounds I could hear were those generated by my body organs. Often things could be somewhat deceiving on the barren grounds. It was hard to tell how deep the meltwater was or how solid or how deep patches or remaining snow were. One day I stepped into what I thought was a shallow snowbank and found myself stuck up to my shoulders in a snow.

Before I knew it, on a bright sunny day, close to the end of July, we packed our personal gear and boarded a Twin Otter airplane to fly back to Resolute Bay. This time, we encountered no fog in Resolute and managed to fly south to Montreal the following day. I thoroughly enjoyed my Arctic excursion and will always remember the Arctic summer climate along with the vastness of the place.

Back to work at the museum, I learned that Terry had set up, with a local publisher, for me to write a book on my northern experiences. Perhaps this was one offer I should have passed up. I was too inexperienced to accurately illustrate this sort of publication with birds and mammals. However, within a few months of my return from the barren grounds of the high Arctic, I managed to write and illustrate a small book, *Arctic Journey*. I am not all that proud of this project. However, another learning experience.

Soon after I got back to the ROM from my arctic adventures, Dr. Scott was getting ready to leave the ROM for good. He soon moved permanently to St. Andrews, New Brunswick. He wanted me to come with him but I decided to stay at the ROM. Before W. B. left the ROM, the board of governors wanted Dr. Scott to become the director of the Royal Ontario Museum. He wanted no part of this and instead he took the director's position at the Huntsman Marine Laboratory in St. Andrews.

By this time the ROM was going through a major expansion project. Some parts of the old ROM structures were demolished. This included the old ichthyology and herpetology department and the old ROM exhibition hall above it. In its place, a new curatorial center and some new gallery spaces were eventually build. I was allowed to join a planning committee, which met once a week, and under the guidance of an outside consulting company aimed to develop some formal procedures on how to design and construct the new ROM galleries and temporary exhibits. Because of my involvement in this committee, I gained some firsthand information regarding these new procedures and eventually became a working partner in the newly established 'exhibit/design' department. With the support of Dr. Scott and from our new director of the museum, Dr. Jim Cruise, I was appointed to a new position organizing and eventually running the art, photography and taxidermy sections of this new department.

I exercised my final responsibility in the Department of Ichthyology and Herpetology and due to the destruction of the ROM's center wing, I organized the physical move of the department from the ROM on Bloor Street and Avenue Road into a converted warehouse on Queen Street West, in the southern part of Toronto. This proved to be quite a challenge. With the assistance of our two summer students, Peter Lori and Paul Witham, along with a couple of other summer students, we managed to pack thousands of bottles of preserved fish, amphibians and reptiles into sturdy boxes and had them moved by a moving company to the temporary location on Queen Street. If I remember correctly, it took all of July and August, and part of September to complete this rather complicated move. Once everything landed in the Queen Street building we reassembled all office furniture and the bottled collections in the new designated spaces. The most challenging and time-consuming work went into the reassembly of the fish, reptile, and amphibian collections. We reassembled all storage shelves in a sizable room and put all of the bottled collection back into a usable order.

It was a definite plus to have the reliable work of the good-humored summer students, who assisted us in this gigantic undertaking. One of their more hilarious contributions occurred shortly after arriving at our Queen Street location. Across the street from the newly located ichthyology and herpetology department was an outfit that prepared live chickens for human consumption. One day one of those live chickens got loose and proudly strolled along the northern sidewalk of Queen Street. As soon as the boys saw the wandering chicken, they decided to capture the white bird and consequently locked it into E. J.'s newly decorated office. When E. J. arrived a couple of hours later and entered his office, he yelled, "Where did the chicken come from?!"

In unison, the boys working in the new collection room yelled back, "From an egg!" By this time the chicken had, of course, done its 'thing' all over E. J.'s office including his desk. We all thought that he would be upset having to clean up the chicken deposits, however, he too thought it all to be very funny and all ended up in good humor and with much laughter.

By the time we completed the Ichthyological move to Queen Street, I was ready to start working in the new exhibit design department.

In the beginning, I was quite skeptical about my move to a totally new department. My stay in the Department of Ichthyology and Herpetology had been enjoyable and very rewarding. I knew that I would miss many of the old ways and most of all I would miss the people. However, things within the museum had changed dramatically. For a few weeks after taking on in my new position, I made daily trips to Queen Street, to assist in further organizing the department and to show my replacement some of the department's procedures.

Shortly before I moved to my new position, ROM senior management had hired the head to run the new exhibit/design department. This person was Lorne Render, a man I had met several years before, at the Glenbow Museum, in Calgary when Terry Shortt and I were researching the Animals in Art exhibition. Again I was very fortunate. Lorne turned out to be a great boss to work for. He was supportive and encouraging in all I was to accomplish. Lorne organized and hired all of the required staff for the building of all new ROM exhibits. The new department was organized into several sections. Each of these was to contribute specific expertise to the design and construction processes of new galleries and temporary exhibits. One of the sections employed interpreters. These people were to communicate with the curators and interpreting their visions for the production of any new public gallery. Another new section consisted of designers who took the interpreters/curatorial explanations and developed design blueprints for each new exhibit project. Finally, my section was to produce all photographs, artworks, and taxidermy productions for any new projects.

Lorne Render was also in charge of the ROM carpenter shop and for the museum preparators. The latter installed all artifacts into their final display positions.

Once many of the new staff had been hired for all of the above sections, the real work began. I was given Terry Shortt's old studio and set up shop immediately. Working on Terry's old worktables, his easel and desk, felt at first a bit strange and it took some time to get used to my new responsibilities and surroundings.

In the meantime, the ROM had become unionized. This changed the sort of staff the institution attracted. Before the union, most ROM staff worked for the institution because they had a passion for the subjects in which they were involved. Most sure as hell did not work at the ROM for the money. However, once the union took hold many new staff members seemed not to have this passion for their subjects. Having said that, however, some of the newcomers did enjoy the challenges of their new positions and soon contributed positively to their designated work in the exhibit/design department.

When I took on my new position, two of Terry's staff members remained in the art section. Both, I think, had applied for my position. One of those staff members was Dave Pepper, who was a multi-talented artist and an excellent craftsman who had successfully accomplished numerous projects in the ROM display areas. Dave had a passion for everything Japanese and often worked on related Japanese projects. I liked Dave and was sad to see him depart from the ROM soon after my arrival. The other staff member left from Terry's days shall remain nameless. It soon became obvious that this person, because of his failure to attain my position, displayed a fair bit of resentment towards me. He, however, remained in the art section for some time before finally departing the ROM. He eventually left the museum when he discovered that I was not interested in office politics and that I would not give in to any of his attempts to dislodge me.

A hell of a way to start a new job. However, I continued to fill the positions I required. The first person I hired was a three-dimensional artist, by the name of Georgia Gunther. Georgia had recently graduated from the Ontario College of Art, and had apparently worked for the art department during her school vacations; contributing positively to the old department. After reviewing her previous accomplishments and her positive attitude for the ROM I offered her the position of three-dimensional artist.

The next artist I hired was Anker Odum. Years before I had through a friend of mine, set up Anker to write and publish numerous articles for the *Reader's Digest* magazine. Anker was offered to write a book for the Digest Company and at the time decided to leave the ROM. However, once 'Ank' had completed the book for the Digest he was looking for a permanent job. With all of Anker's talents and experience, I was happy to have him as an integral part of my section.

A lot of resentment from the one staff member left over from Terry Shortt's days continued. I pretty much ignored all of the office politics and with the positive backing of Lorne Render continued to do my thing. I was, however, surprised and saddened to learn that Anker was very much a part of a conspiracy to get rid of me. I could never quite figure out what he was thinking.

Much of the old museum ways had changed. I always enjoyed associating with a lot of the ROM staff members and kept myself informed of ongoing activities. Suddenly, all of those associations were gone. I was not invited to any staff breaks or any conversations regarding our work procedures unless I called a formal meeting. I found myself to be pretty much on my own and realized that those who were in charge of things were not welcome to mingle with the workers. Welcome to management!

Just the same, I continued to pursue my responsibilities. I still had to hire one more staff member to complete the required personnel in my section. I had to find a qualified museum taxidermist. This did not prove to be easy. The ROM had not had a Taxidermist on staff for many years. Lorne Render and I even traveled to the Milwaukee

Museum, Wisconsin, in the USA to see if we could persuade their taxidermist to join the ROM. This museum had a qualified taxidermist on staff who was apparently interested in moving to another institution. Just when we thought that we had convinced the Milwaukee Museum taxidermist to come and work for the ROM, he decided to stay in Milwaukee.

Our visit to the Milwaukee Museum, however, turned out not to be a total waste. I learned various procedures on how to make realistic looking support artifacts for many life science displays. One of these was how to produce artificial, realistic looking vegetation including how to make artificial tree leaves. These were made out of plaster molds made from freshly picked leaves. Inserting thin sheets of plastic into the plaster leaf molds and placing this combination into a vacuum machine producing plastic leaves accurately resembling the original leaves. These had to be cut out of the vacuum formed plastic sheets and then had to be hand-painted to resemble the original leaves. It was quite a time-consuming process. However, it really produced worthwhile and long-lasting support items which added much depth to many natural history displays.

For quite some time after, Lorne and I had visited the Milwaukee Museum; I continued to advertise in local newspapers and magazines searching for a museum taxidermist. This too proved to be unsuccessful. After searching for several months, I was ready to give up, when one day, quite unexpectedly, a young man walked in off the street and showed me pictures of some birds and mammals he had mounted. I was impressed. I knew that with some additional training in the art of museum taxidermy, this young man had real potential. The young man was Peter Knapton, whom I decided to hire. We soon set up a taxidermy shop and I sent Peter to a few American natural history institutions and a couple of well-known established taxidermists in the USA. He learned quickly and soon produced museum quality work.

Finally, all of my required staff members were on board. I did not have to hire any new staff for the photography area. All staff had been previously hired by Lee Warren the former head of the ROM photography department. Two photographers: a darkroom technician and a secretary remained in this section. All of the staff members in the photography section were well established and well qualified. Most of the ROM photography required was for the ROM curatorial departments, taking and developing black and white reference pictures of artifacts. Only occasionally where their expertise required for gallery projects.

By now, plans for the upcoming museum expansion were in full swing. These plans often required that the exhibit/design sections had to move from one area to another and we often found ourselves working in cramped quarters. One of the more drastic and lengthy moves was an office building, across the street from the ROM, on Bloor Street next to the Park Plaza hotel. During this time, the taxidermy shop and the ROM carpenter shop were also moved off-site to Queen Street into the same building into which we had moved the ichthyology and herpetology department.

At our temporary location on Bloor Street, we were crowded into two rooms and continued to interpret, design, and produce artwork for the new galleries. We spent more than a year in the office building on Bloor Street before moving back to the ROM into our permanent work location. Back at the ROM, the exhibit/design department was located in the lower floor of the old museum's southwest wing. All sections associated with the exhibit design department: the carpenter shop, the preparators section, the photography section, the taxidermy shop, the design section, and the art section occupied the total area of the lower floor. We now had sufficient workspace and many of us had large windows facing Philosopher's walk of the U. of T. campus. Unfortunately, the taxidermy shop had no windows and was now located in two large rooms just to the north of the art section on the east side of the long exhibit/design hallway. Just down from the art section, Lorne Render's office located almost directly across from the art section, on the east side of the exhibit/design hallway was the photography section. The photography section was a good-sized space, which included office space, a large photographic studio, and a darkroom. The photography studio also had no windows for good reason. Each artifact that had to be photographed had to be lit individually to accomplish the desired effect for each project.

From the aforementioned areas, we started to pursue all that was asked of us. We soon developed art, some photography, and taxidermy for new galleries. For a while, things were going well. Everybody had settled into their designated positions and everyone seemed to be enjoying their work. We tried hard to work with the interpreters and designers to produce whatever was requested of us. Unfortunately, this was not always easy. Some of the curatorial staff did not enjoy working with the interpreters or with the designers. Many curators were totally unaccustomed to the new procedures and processes. This was especially true with some of the life science curators many of whom I got to know rather well during my stay in the ichthyology and herpetology department. They subsequently felt more comfortable working with me directly, often causing problems, upsetting interpreters and designers who accused me of interfering with their responsibilities. I could hardly blame them for their feelings; however, I was caught in the middle and could rarely explain my awkward situation to them. At times it was difficult to work in harmony in the exhibit design department no matter how hard I tried to explain some of these undetermined situations. Nevertheless, we always managed to muddle through and accomplish satisfactory results in the end.

One of the first substantial galleries to be built was a bat cave. The senior curator, Dr. Randolph Peterson, of the ROM mammalogy department, was a world expert on bats and eagerly wanted to display some of his expertise in a ROM gallery. He wanted this project to be unique and to have a lasting and informative impression on ROM visitors. Dr. Peterson suggested we use the St. Clair Cave, located in Jamaica as our prototype. St. Clair Cave is a large cave, several miles long; located in the Jamaican mountains just outside of Ewarton?

To make sure that the above location was feasible as a prototype for a ROM bat cave display, we made an investigatory trip to the location. I invited Dr. Brock Fenton, whom I had gotten to know while working in the ichthyology department to accompany me on this trip to the Jamaican cave. Brock and I flew to Kingston, Jamaica; rented an old American car at the airport and drove to Kingston and checked into a gated hotel. The following morning, in our old right-hand drive car, we set off for Ewerton, a small town in the middle of the Jamaican Island. Driving on the left-hand side of the road proved to be most uncomfortable for me and we soon decided that Brock should be the driver in Jamaica. Once we reached Ewerton, we took a small dirt road off the main highway; eventually reaching a small farming community some two or three miles from the St. Clair Cave location. In this small community, we hired a young man to lead us to the cave. It turned out to be a fair hike to the cave's sinkhole. We passed through numerous marijuana fields, crossed a river, and halfway up a slippery, narrow mountain path reached the large, deep sinkhole of St. Clair Cave. On the way to the cave, I admired large clusters of bamboo stands. I had never seen such large ones before.

Brock was well prepared. Once we reached the deep sinkhole of St. Clair Cave, Brock took out of his backpack a lengthy rope ladder which we tied to the bottom of a nearby tree at the top of the sinkhole and we slowly descended down the swaying ladder to the bottom of the rocky sinkhole. The steps in the rope ladder proved to be a bit narrow for the width of my feet and I often got one or the other foot temporarily stuck in one of the runs. We did, however, reach the bottom of the sinkhole safely, tied on headlamps, crawled over numerous large boulders, and entered the totally dark cavern. With the headlamps turned off, you could see absolutely nothing, not even your hand in front of your face. We walked through several large caverns with their ceilings and floors covered in all sorts of large, variously shaped stalactites and stalagmites. The ground of the cave was very uneven, slippery and progress was often slow. Further on into the cave in some large openings, the ceilings were covered with thousands of bats. According to Brock, these dense, large clusters of bats consisted of several species, all of which, each evening, exited the cave to forage across the Jamaican countryside. We took pictures of the cave walls, ceilings and floor and I formed mental impressions of sections I thought suitable for smaller reproductions back at the ROM. Brock and I spent most of the day in St. Clair Cave, documenting many aspects of the large cavern, including the sinkhole. At dusk thousands of bats did indeed exit the cave, flying out in great swarms into the darkening landscape.

It was late that evening when we drove back to the hotel in Kingston. We rode the hotel elevator to our room on the third floor. Some of the other hotel guests did not appreciate our presence in the elevator because we were covered in mud and bat guano, and most likely did not smell good. To add to our dismal appearance, Brock had captured several live bats swirling in a cloth bag which he was carrying in one hand. He subsequently released the bats to fly around our hotel room. All of these bats were taken back to the ROM for proper identification and to be shown to the members of senior management and to the curatorial staff of the mammalogy department.

Following our brief investigatory cave trip to Jamaica, all those involved in the bat cave project back at the ROM became excited about building the ROM cave. In the meantime, gallery spaces had been designated and the construction of cave was approved. I chose staff to accompany Brock Fenton and me to go back to Jamaica to gather all of the required documentation and specimens to build the ROM cave. We purchased the necessary equipment for this major collecting expedition and soon found ourselves back in Jamaica.

Before leaving, I had contacted the CBC television network to see if they were interested in filming our Jamaican bat cave excursion. Surprisingly, they were interested and several CBC staff accompanied us to the St. Clair Cave exploration. The CBC made a documentary film of the Jamaican cave collecting trip and eventually aired a film as part of *The Nature of Things* television series.

The ROM staff I chose to accompany Brock Fenton and me to document all aspects of the cave was a ROM photographer, two taxidermists, one artist, and Bob Barnett, the lead man of our ROM interpretive section. Bob was very helpful and was kind enough to look after all of the financial requirements for and during our stay in Jamaica. Bob also assisted in collecting numerous cave invertebrates and a variety of other cave specimens.

Once we established a home base in Kingston, Jamaica we drove each morning from Kingston to Ewerton. From there, we drove along the dirt road leading to the small farming community were we hired a couple of young men to assist us to carry all of our gear to the St. Clair Cave sinkhole. Each morning I enjoyed the trek through lengthy grass-covered fields, sections of which were crowded with marijuana plants and a few other crops. Bright green-colored birds, resembling some sort of large parakeet and a variety of hummingbirds were frequently flying about. By the time we had made our way past the open fields and crossed the flowing river, we reached the bottom of the mountain trail leading to the cave's sinkhole. Once again, I noticed on the riverbanks substantial stands of tall and thick-stemmed bamboo. After crossing slippery river rocks we made our way up the narrow muddy trail up the mountainside to the cave's sinkhole. On the way up the trail, patches of morning sunlight filtering through the trees illuminating some low growing coffee plants many decorated with bright red coffee berries.

At this time I must mention a rather intriguing encounter which occurred one morning on the way to the cave. While passing through one of the marijuana fields out of nowhere, a young lady appeared. She walked up to Anker and propositioned him. Anker stopped, scratched his head, and answered without much hesitation, "Maybe on the way back." I never understood why the young lady propositioned Anker. Perhaps because he was the slightest and smallest guy in our group or because he was lagging somewhat behind us.

By the time we reached the cave's sinkhole morning temperatures had risen substantially. Reaching the sinkhole after the first morning's trek to the cave, I discovered to my demise that some of my staff refused to enter the cave. I was disappointed in their decision and asked myself why in hell had they volunteered to come to Jamaica? They had been told, back at the museum, what cave collecting was all about. All of a sudden, they were afraid of contracting histoplasmosis, a lung infection caused by breathing in spores of fungus found in bat guano. In retrospect, I guess these people were smarter than I was. Quite some time after our cave expedition, when I went to my doctor for my annual checkup, an x-ray revealed that I had a problem with my left lung. After a major lung surgery, it was discovered that I had scar tissue on my left lung that was caused by histoplasmosis.

Anyhow, back to our cave explorations. The interior of the cave was tremendously hot and humid. Everyone entering the cave was told to wear some sort of cloth over their nose and mouth while working in the cavern. I tried this but found it difficult to breathe while collecting in the cavern.

Our photographer, Bill Roberson, did a fantastic job photographing all aspects of the cave from which we could reproduce an interpretive smaller version of St. Clair Cave back at the ROM. Anker spent most of his time sketching and painting detailed studies of the cave's sinkhole. These proved to be excellent and made terrific references for the large painting of the sinkhole he eventually painted back at the ROM.

Each evening, just before sunset, we set mist nets at the entrance of the cavern and collected bats of each species some of which were taken back to the ROM. We made molds of these to reproduce the many bats required for our museum cave. We did not want to kill too many bats and thus ended up making hundreds of accurate wax bat models for display in the museum cave.

We spent about two weeks in Jamaica, collecting and documenting as much as possible of the St. Clair Cave structures. All of these were used in the ROM cave replica.

The Jamaican collecting trip turned out to be most successful and we all managed to return to Toronto in one piece. We did, however, encounter some problems with the Canadian customs people at the Toronto Pearson airport. Apparently, the airport security dogs went absolutely wild when introduced to our suitcases and to our collected specimens and artifacts. It took some time before a customs officer finally came and talked to us. He asked, "What in hell were you all doing in Jamaica?" After explaining to the officers that we had been collecting flora and fauna for the construction of a Royal Ontario Museum bat cave and in order to do so we had to daily walk through numerous marijuana fields, they understood why their search dogs were so interested in our belongings and they soon allowed us to enter back into Canada.

Back at the ROM, I was excited about building our bat cave. Unfortunately, those who I had chosen to go on the Jamaican trip with me showed little interest in participating and I was pretty much left on my own to design and to build the ROM cave. The unionized staff had more rights than ever and I decided that it was not worth the aggravation to challenge their attitude. I soon built a four-by-eight-foot model of what I thought the ROM cave interpretation should look like. This model was reviewed by the ROM mammalogy curatorial staff and was approved after a few minor changes.

Once the cave model was completed, I had to hire some contract staff to assist in the construction of the museum cave. I hired a few more artisans whose time was to be totally dedicated to the cave construction. A ROM staff designer designated for the project to provide us with blueprints, based on the model, for the cave workers to work from. Unfortunately, even the detailed scale model of the cave was not a sufficient reference to accomplish this. The designer assigned to us turned out to be useless. She instructed the carpenters to install the metal lath foundations, representing cave stalactites and stalagmites, horizontally instead of vertically. I'm not sure whether or not she ever revisited the cave model. One would think that any full-grown person would have read about or seen some pictures of how stalactites and stalagmites are formed – apparently not! The lath had to be torn out and reinstalled. To top it all off, the designer accused me from that day one, and for years to come, of badmouthing her to the carpenters about her screw up. She had decided that it was part of her job to complain about me to all who would listen. I did, of course, talk to the carpenters to sort out the problem. Following this dismal experience, I decided not to use her services any further.

I experienced some continuous staff problems throughout the cave construction. Once the required three-month probation periods were up, their work output slowed significantly and no matter how many meetings we had they frequently deciding to proceed with their own construction procedures. Because of these self-serving decisions, work frequently slowed down and many times things had to be done twice. Their needless interventions soon caused us to go over the budget.

During these times, the head curator of the mammalogy department called to inform me that he had a sponsor who was interested in donating some funds to the project. *Boy,* I thought, *we could use that kind of support right about now.*

The cave had a budget of one hundred thousand dollars to pay for fieldwork, all materials and all staff time required to build the cave. It was the latter which caused all financial overruns. I now hoped that the curator's donor would be of real financial assistance. As it turned out, the donor donated two thousand dollars, which in reality did not make all that much of a difference.

In the end, I cast wax bats on my own time just to get the damn job completed. I also took turns spraying liquid plaster on the metal lath foundation surfaces. This turned out to be the best and fastest way to cover all of the metal lath foundations. This was quite a messy job and anyone operating the plaster sprayer ended up, from head to toe, covered in plaster. By this time, I was ready to do anything to complete the project. Over budget and over the time allotted, I kept on struggling to complete the cave. All of my initial enthusiasm was slowly disappearing and I silently hoped that this was not an indication of how all new gallery projects would proceed.

Once the cave project was finally completed, I was pleased with the final result and with what we had accomplished under difficult circumstances. I guess the saving grace of all of the trials and tribulations was that the ROM bat cave was well received by the public. The head curator of mammalogy, Dr. Peterson, was more than pleased and some forty years after completion, the cave still attracts thousands of visitors.

In the meantime, while all of our gallery planning and construction continued. The artists and taxidermists in my section proceeded to work on all sorts of projects. Anker produced many illustrations for curatorial publication and for hundreds of gallery labels. All of the old live science display kept from the old galleries was moved from the third floor of the museum to their new locations on the second floor. The reptile gallery was taken apart but all of the reptile casts/reproductions were saved. Hundreds of bird mounts were destroyed and most of them were thrown out. Most mammal mounts went into storage in the mammalogy department. The fish gallery was taken apart and all cast and mounted specimens were also put into storage. Two of the oldest and most historical dioramas were also dismantled, never to be seen again. One of these was of Ontario black bears. The other was the passenger pigeon diorama which I think should definitely not have been taken apart and should have become a historical part of the new bird gallery. The background painting of this diorama was expertly painted representing the natural surroundings of the Humber River just west of Toronto when passenger pigeons were still plentiful. The background painting included hundreds of passenger pigeon both in flight and resting in trees. This site is gone forever. It should have never been destroyed and saved to be experienced by future generations. The last live passenger pigeon, named Martha, died at the Detroit Zoo in 1912.

While all of the destruction and relocation of the old life science displays was going on, we made plans for the new mammalogy gallery. At the same time, we were asked to construct two major dioramas for the new New World Archaeology gallery in the basement of the museum. One of these dioramas was to display North American native people who had hunted and killed a young mammoth. We created, a life-sized, recently slaughtered young mammoth being skinned by two native people. For the construction of this diorama, I hired two sister artists: Linda and Sandra Shaw. I had met both of these artists some time before and Linda had just completed several illustrative art projects for the mammalogy department. Both of these artists were imaginative and very competent workers.

To construct a dead mammoth was challenging to say the least; however, I was very fortunate to have the Shaw sisters on staff to produce the realistic and convincing artifacts required for the project. Linda was put in charge of building the mammoth and Sandra was charged with interpreting the required postures and constructing several life-sized figures of native people. Both turned out to be of superb quality and added much to the dioramas.

After putting our heads together, we soon decided the sort of materials required to build the exhibit for the New World Archaeology gallery. For the mammoth, we decided to use Highland cattle hides to mimic mammoth fur. We located a local Highland cattle farmer who was willing to sell us two live Highland cows. We transported the cattle to a local slaughterhouse for processing. The farm from which we bought our cows resembled many typical southern Ontario farms. The Victorian farmhouse had a substantial barn and acres of healthy grassy fields in which a small herd of shaggy Highland cattle grazed.

To acquire our needed cowhides, Bob Barnett, the head of our interpretive section, was kind enough to accompany me to the cattle farm and assist in transporting our purchased cows to a local slaughterhouse. Watching the procedures to put the cattle down at the slaughterhouse was not exactly a pleasant experience. However, it had to be done.

Arriving at the slaughterhouse with our two live highland cows, the slaughterhouse staff put both cattle down, removed their hides, and cut cow bodies into edible portions for human consumption. The meat of the Highland cows was very lean, pretty much without any visible fat. Back at the ROM, I handed out portions of our expertly cut meat proportions to all who were involved in the New World Archaeology project.

The cattle hides were sent out to be tanned before the construction of our mammoth began. At the same time, Linda visited the main Toronto slaughterhouse and managed to get, from a freshly killed cow, the exterior structures surrounding a cow's eye. This she preserved and used in the eye construction for her creations of the young mammoth. Linda did indeed create a damn good authentic-looking young mammoth. Both of the Shaw sisters created excellent and authentic looking artifacts for the New World Archaeology gallery.

I added one more artist to complete the above projects. This artist had previously been working on the bat cave and continued to change previously worked out procedures and again took more time than needed to accomplish the archaeology project. One day, while checking on the progress of the New World Archaeology gallery I found him sitting on the floor in the gallery space mixing all sorts of materials. When I asked him what he was doing, he answered, "I'm making dirt for the diorama." I could not believe it! I told him that it would be much more realistic and definitely less time consuming to go out and buy bags of soil from a local nursery and run this through our fumigation chamber and install this wherever it was needed.

For the archaeology dioramas, I also hired an artist to paint the diorama background scenes. She too did a decent job and her paintings were eventually photographed and enlarged to cover and the curved diorama walls. Unfortunately, this particular artist later on in her employment caused more union problems for me. She became romantically involved with one of our other art section employees and together decided that they could run my section better than I could. They soon made this known in the art section and even more so around the whole museum. They insisted on continuously spreading false rumors and worked hard at office politics causing continuous problems. I finally had to fire the boyfriend. Of course, this immediately got the union involved and wasted much time and continued on for several months. Eventually, the problem was resolved at arbitration and the young man was dismissed.

I guess these sort of staff problems are quite common in other institutions. I admit I pushed workers to get jobs done on time and within budget. I think that working in any government institution time and money are of no consequence. I often experienced this sort of thinking when buying equipment from outside suppliers. Many times, when any given supplier found out who the ordered materials were for, costs would suddenly rise substantially.

In the meantime, the production of the archaeology dioramas went pretty smoothly. The curator in charge of the gallery was a delightful man. He was very cooperative and supportive; he made our jobs go as smoothly as possible.

All artists involved in the evolution and in the construction of the New World Archaeology dioramas did great work and the completion of the project resulted in accurate and well-interpreted productions.

While working on the New World Archaeology gallery the infrastructures for mammalogy gallery were being built. After the dismal experience we had with the designer on the bat cave, we decided to go without one in the preparation of the rest of the new mammalogy gallery. Working without a designer, of course, created more union problems and some days I found more yellow union complaint sheets in my mailbox than mail. These, however, with the support Lorne Render, soon vanished. All proceeded fairly well in the design of the mammalogy gallery and construction soon followed.

Before we started the actual construction on the new ROM mammalogy gallery, Bob Barnett and I decided that we should visit the mammal gallery displays at the Natural History Museum in Victoria, British Columbia. This turned out to be worthwhile. I found their use of different ceiling heights throughout their galleries identifying display spaces from one area to another very useful. I was impressed with their construction and display of a full-sized mammoth in the center of their mammal gallery surrounded by various other mammal displays. Their mammal mounts were excellent. The fish gallery at the British Columbia museum was also well done. Cast of various endemic species and the finishes of these were impressive.

Returning from B.C. and following some lengthy meetings with the head of the ROM mammalogy department, we decided to create a series of realistic

Canadian habitats in the form of full-sized dioramas for the new gallery. The entrance to our mammal gallery was to display an open Ontario hardwood forest scene in full fall colors. The background of this open diorama was constructed in two parts. A rock wall fading into a hand-painted scene, portraying a typical southern Ontario hardwood forest. For the diorama background painting, I hired a young, very competent artist by the name of Dwayne Hardy. Dwayne had studied with Clarence Tillenius, a well-known and experienced museum artist working out of Winnipeg, Manitoba. He also studied his craft with another Canadian artist who had moved to the USA some years before working out of New Mexico. Dwayne had studied with and learned a lot about 'plain air' painting with this master and was capable of painting exquisite, realistic museum diorama backgrounds very convincingly. For the Ontario hardwood forest diorama Dwayne spent time in Algonquin Park and in the forest of the Parry Sound area. He carried out painting studies of various habitats and documented all he required to complete the ROM hardwood forest diorama.

Peter Knapton, our head taxidermist, mounted a group of white-tailed deer to be the main attraction in our open hardwood forest diorama. Peter also mounted some smaller Ontario mammals, including a red fox for this display.

We wanted the visiting public to be a part of our hardwood forest display. To accomplish this we produced in fiberglass, several full-grown realistic looking maple trees standing both inside and outside of our hardwood diorama. We made latex rubber molds of full-grown maple tree trunks in the Ontario Halliburton Highlands. Two of my artists spent a week painting layers and layers of liquid latex rubber, reinforced with sheets rubber coated cheesecloth creating negative molds of several large maple trees. These molds I farmed out to a ROM staff member by the name of Gord Gyrmov, the senior technician in the ROM's paleontology department. Gord really knew what he was doing and was an expert in casting anything in fiberglass. Gord created, out of fiberglass, several large, accurate full-sized maple tree trunks which reached from floor to ceiling in the display areas. Once these tree trunks had been hand-painted to resemble their living counterparts, it was time to install lots of maple leaf covered branches. We collected many maple tree branches of various length and thicknesses which we dried, fumigated and eventually were covered in artificial colorful maple leaves and inserted these branches into the fiberglass tree trunks.

I had to acquire thousands of artificial realistic looking maple leaves. I tried to buy these locally and discovered that these were quite expensive. With the assistance of the manager of the ROM gift shop, I made contact with a company in Hong Kong who were able to, for a fraction of the local prices, accurately reproduce colorful, artificial maple leaves from the freshly picked leaf samples sent to them. I ended up traveling to Hong Kong to personally observe the production of the many required leaves for our open forest diorama.

My trip to Hong Kong was long and tedious. However, once there I soon found it to be a most inviting and interesting place. People were friendly. I was able to travel by myself without any problems anywhere within the city. The Hong Kong subway system was efficient and the shiny inside and outside stainless steel subway carriages made it easy to get around. I very much enjoyed the food in local restaurants. One of the more interesting places I ran into in this very busy city was a street called 'Bird Street.' No cameras were allowed on this street. Both sides of Bird Street were lined with thousands of caged birds of all sorts and sizes, all were for sale. Most, if not all of these birds, I'm sure, were illegal.

I spent the first day in Hong Kong familiarizing myself with the city and I got to know my suppliers. The following day, I was invited by the manager/owner of the company to travel by train to the Hong Kong territories some distance outside of the city. Before boarding the train, the company owner invited me for lunch at the train station leading to the territories. We sat in the largest restaurant I had ever experienced. The waiters in this establishment came around pushing large carts, loaded with food of all description from which we could pick whatever we desired. We ended up having the sort of lunch I would not soon forget.

Following our inspiring lunch, we boarded our train and less than an hour later arrived in the Hong Kong territories. There, we met people who were experts in producing artificial artifacts of all descriptions. After practicing the Hong Kong tradition of exchanging business cards, we discussed prices and production time, and I soon placed an order for several thousand artificial maple leaves.

We ended up with a bit of spare time waiting for our return train to Hong Kong. I found walking around the territory most interesting. I discovered that at the entrance of several homes, jars of human bones were on display. These bones, I was told, were the remains of ancestors.

Although I found this a bit weird I respected the traditions of their culture.

Leaf mission accomplished and back in Hong Kong I soon flew via Vancouver, back to Toronto. It might be hard to believe, but the trip to Hong Kong for our artificial leaf production saved us a lot of money.

Amazingly, not long after returning to the ROM the shipment of artificial, colorful maple leaves arrived. Now the arduous task of attaching these one by one, to our tree branches. Each leaf was tied onto the branches with floral tape, a very time-consuming and tedious job. Once this task was accomplished, all the leaf-covered branches were carefully inserted and glued into the upper parts of our fiberglass tree trunks. The branches had to be carefully positioned into the tree trunks to resemble natural growth.

I always took great pride in working within budgets and completing all jobs on time. However, by this time I became convinced that this was impossible when working with union labor in a museum.

By the time we had completed the hardwood forest diorama, it was time to start working on the rest of the mammalogy gallery. Several diorama shells were constructed and were ready to be painted and filled with the flora and fauna of various Canadian habitats.

I always wanted to paint and build a diorama myself. My boss, Lorne Render, was kind enough to allow me to do so. This was to be an after hour project and I would spend weekends and work after hours to accomplish my project. I chose to do the smallest of the planned mammalogy dioramas. It was a tall, narrow diorama which was to interpret the Blacker Wood Forest, located in southern

Ontario near Lake Erie. Here, tulip and gum trees grow which do not grow anywhere else in Canada. I made full-day trips to the Ontario Blacker Wood Forest, photographed and made some full color sketches of the vegetation found in that inhabit. I collected leaves and any other natural materials useful for the production of my planned diorama. The scene I was to create was a springtime forest with dogwood bushes and white trilliums in full bloom. Later, I would have to return to the location to make latex molds of some of the required trees and make plaster molds of their leaves to create accurate vacuum formed replicas for the tree's foliage. All of the foliage had to be vacuum formed and had to be hand-painted to create realistic looking artifacts. Again, this was a tedious and time-consuming job to be accomplished down the road.

Later, I would have one of our taxidermists collect and mount an Ontario gray fox, which was to be the star of my diorama. The Ontario gray fox is somewhat unique, it is capable of climbing trees – the only fox to do so. A few other small indigenous mammals to the Backer Woods area would also have to be collected and prepared to become part of the Blacker Wood habitat group.

I painted a large study of the Blacker Woods forest which had to be approved by the mammalogy curatorial staff and which I then could use as a reference for the final diorama backdrop painting.

While working on the Blacker Woods diorama, the head curator of mammalogy wanted us to start working on the rest of the planned dioramas for the gallery. One of these was to create a mountain meadow. To accomplish this I arranged with a group of western cowboys to lead us into the Alberta Mountains close to the British Columbian border. I also had to hire a couple of new artists to accompany us on the mountain trip and to eventually paint the background and install all of the flora and fauna for this project. I took a ROM photographer along to document all of the required scenery and a taxidermist to collect and to prepare the collected specimens to take back to the ROM. Near the Alberta Montana border, we traveled into the mountains with all of our gear loaded into the backs of two pickup trucks. We entered the mountains near the Montana, Alberta border and drove slowly for several miles over a rough dirt road into the high country. When we reached the location the cowboys had picked for us, we set up camp in a valley surrounded by mountains – truly breathtaking. We soon set up several tents to sleep in and one large tent to cook and to eat in.

In the valley and in the surrounding mountains we managed to collect several small mammals which included marmots, pika, and some ground squirrels. The pika were to me, the most interesting of the small mammals in the area. I had never seen these rather elusive small mountain mammals in life before. It took us some time to locate the pikas in the rocky rubble about halfway up a mountainside. Their skittish and fast movements made them difficult to see and even more difficult to collect. All of the collected mammals were skinned and prepared to be later mounted for the exhibit back at the ROM. We also collected a fair bit of vegetation to be accurately reproduced back at the museum. Higher up in the mountains we encountered some bighorn sheep which, fortunately, we did not have to collect. This was grizzly bear country and when traveling through tall grasses or in any waist-high vegetation, I always carried a rifle with me for protection. I only spent a week in the alpine meadow site. I had to get back to the ROM to participate in another upcoming field trip.

Arriving back at the museum from the mountains, I received a telegram from our cowboy guides informing me that they were holding my staff for ransom. My reply was short and sweet – "Keep them!"

The ROM staff and all of the collected flora and fauna from the alpine meadow location arrived at the ROM about two weeks after I had departed the site. We were now ready to fabricate the alpine meadow diorama. The two artists I had hired prepared the curved back wall of the diorama and began to paint a mountain scene. Unfortunately, not long after they started I was informed by management that they had run out of money and we had to stop working on the mammalogy gallery. Nevertheless, I had to pay the two artists hired for this project to honor their contract with the ROM. We were obliged to pay them the full amount stipulated in their contract with the ROM. The end of the mammalogy gallery production also included the cancellation of my Blacker Woods diorama and consequently I never got to build a diorama on my own.

The next big project I became involved in was to construct an Australian desert reptile gallery. I guessed that the funds for this project had been set aside and were available. This turned out to be a once in a lifetime museum expedition. I was to work with two ROM herpetologists in the red center of Australia to collect the necessary flora and fauna for the project. I took Anker Odem with me to Australia to assist me in collecting the necessary materials for this enticing project. We flew from Toronto to Los Angeles, from there to Hawaii, on to Auckland New Zealand and from there to Sydney Australia. We spent a day and a half in Sydney after which we took a four-hour flight north to Alice Springs, located in the red center of Australia. The landscapes in and around Alice Springs were absolutely amazing. The red sandy ground color, the local vegetation, and the colors of the surrounding hill sites were absolutely striking. Reds, oranges, and various shades of purples all highlighted by bright sunshine.

Soon after arriving in Alice Springs we all settled in at the Desert Rose Hotel. The next day, we rented a couple of old cars and we split into two collecting groups. The herpetologists went off to collect whatever reptiles they encountered. Anker and I did pretty much the same. However, our collecting priorities were somewhat different. We too collected some reptiles and spent time making latex molds of termite hills and certain tree trunks. We cut and preserved various grasses and whatever other vegetation we thought we might need for the Australian reptile display back at the ROM. Anker made several full color studies of surrounding sceneries which would be useful for any larger reproductions at the ROM.

The head herpetologist accompanying us, Dr. Bob Murphy was a Vietnam veteran. Bob was the only person I ever met who did not mind being stationed in Vietnam with the US armed forces. Apparently, while on duty in the US army in Vietnam he spent much of his time collecting amphibians and reptiles. Bob had a great sense of humor which he adapted to most interventions making him a pleasure to work with. Before we engaged in our collecting tours in the areas around Alice Springs we connected with a local naturalist who was kind enough to show us some of the best collecting sites and how and where to travel to do so. This man was very knowledgeable of the local flora and fauna and was of much assistance to all of us throughout our stay in the Red Center. We also hooked up with personnel from the Alice Spring Museum. The curator of this institution was of much assistance to us and provided us with local traveling information. He also allowed us to use the museum's laboratory to prepare specimens for shipment to Canada. The Alice Springs Museum galleries displayed a comprehensive array of local reptiles. All of these had been freeze-dried in natural poses and then hand-painted to accurately resemble their living counterparts. Freeze-drying reptiles for display purposes was certainly something to think about. However, our ROM taxidermists and artists had no interest in exploring these possibilities.

Anker and I continued to collect whatever we thought would be useful in the construction of the ROM Australian Desert gallery. We collected spinifex, gum tree leaves, blue mally branches and various other shrubs and grasses which we eventually shipped home.

Most days in the desert were extremely hot; temperatures remained in mid-forties. The locals told us that it was a dry heat and that you would not find the temperature all that uncomfortable. Well, let me tell you, for a couple of Canadians it was damn hot!

Each day on our trips into the Red Desert, Anker and I discovered something new. One morning we encountered a large monitor lizard more than three feet long which I really wanted to catch. A cast of such an impressive lizard would have added a lot of punch to our ROM exhibit. I chased the monitor up a gentle red sandy slope covered in spots with bunches of prickly spinifex grasses. I ran as fast as I could, and each time I caught up to the traveling reptile it somehow switched into another gear and without any effort soon ran faster than I could. I never did catch that damn monitor – it was simply too fast for me. I had asked Anker to bring a large bag, in which we could put the lizard, in case I was able to catch the animal. Being out of breath I returned to the bottom of the hill where good old Anker was standing relieving himself. Needless to say, it was not a bag he was holding in his hand. On another occasion while driving along the local highway, we saw in front of us a good-sized lizard sunning itself on the side of the highway. The lizard had no fear of us and I was able to drive the car right up next to it. I told Ank to keep an eye on the lizard while I searched through the car's trunk for some sort of container into which to put the animal once captured. While I was searching through the trunk of the car a sudden, tremendous noise interrupted my search. An Australian 'road train,' a larger tractor-trailer pulling two full-sized other trailers passed us at considerable speed. When I looked up, Anker was still leaning on the fender of the car, still looking at the lizard. When I looked at the reptile on the side of the road, it was still in its original position; however, it was now as flat as a pancake. The road train had run over it. So much for that specimen.

Anker and I continued to make our daily collecting excursion around the Alice Springs area. All the riverbeds in and out of town had completely dried up. The only inhabitants of these dried out riverbeds were groups of native people sitting in the middle of them. One day we did, however, find a small shallow stream on the outskirts of Alice which still had a bit of running water in it. On close examination, I saw a few fish swimming in this small creek. Using a dip net, I managed to catch a few of these and soon identified them to be Australian rainbow fish. I had only ever seen these fish in aquaria at home and got quite a kick seeing them alive in their natural habitat.

Following our fish discovery, we drove further out of town and I suddenly heard, which was to me, quite a familiar sound. It took a few minutes for the sounds to sink in but suddenly I knew what it was. We had encountered a huge flock of zebra finches. There must have been several hundred of these small red-beaked birds flitting about in the local shrubbery. Again, this too was quite a unique experience for me and it made my day!

One day we set out to go collecting with the herpetologists. Not too far out of town walking about in grasses and through some shrubbery, we discovered a good-sized borrow in the ground. Bob Murphy knew immediately what this burrow was all about. He yelled out, "A king brown!" Before I knew it, I was digging up the ground around the hole in the ground, determined to collect this very poisonous snake. Anker stood by smoking a cigarette and this time he was holding a cloth bag in one hand. Bob had a snake snare in hand and was ready to capture the reptile once it emerged from its burrow. It did not take very long for me to dig out the venomous reptile. When the king brown emerged from the ground, Bob immediately tightened the snare behind its head and deposited the vigorously scrambling reptile in the bag Anker was holding.

After capture, we immediately returned to the hotel and deposited the venomous animal into cold storage to humanely put it to sleep. This king brown snake was over three feet long and was the most dangerous animal we collected in Australia.

Each evening, we all met in one of our rooms to compare our day's experiences and to see what each of us had collected. On one of these evenings, Bob Murphy was holding a good-sized lizard, tightly behind its head to show us one of the animals they had collected that day. Anker asked Bob Murphy, "Does it bite?"

Bob answered, "No." Anker put one of his hands to the lizard's mouth and was immediately firmly bitten drawing blood.

One of the striking things we experienced on our daily trips in the Red Desert around Alice Springs were all of the dead animals, most of which had died from lack of water. Dead cows and horses in various stages of decomposition were a common site. Many of these well-decomposed animals now made homes for a variety of living lizards. On one of our morning excursions, Anker and I encountered a huge flock of budgerigars. These landed not far in front of us, taking advantage of one of the few remaining waterholes. The budgerigars, native to Australia, were all green in color. None displayed the variety of colors we are used to seeing on budgies in local pet shops back home. What a site! When finished drinking, this substantial flock of budgerigars, all at the same time, took flight from the waterhole and looked like miniature falcons swooping over the Red Desert. Quite a sight, one I would not soon forget.

Throughout our stay near and in Alice Springs, we saw all sorts of incredible birds. To us, these were all very exotic and many were absolutely awesome. We saw red-tailed black cockatoos, the odd Major Mitchell cockatoos, some princess parrots, little corellas, and numerous crested pigeons. Bird life in the surrounding desert was truly remarkable. It was not uncommon to see flocks of white cockatoos flying directly over us. We saw wedge-tailed eagles, and whistling kites hunting the wide-open spaces of the desert. One morning, we encountered numerous wild rabbit which had been introduced to Australia years before. Although very briefly, we saw a dingo crossing the road directly in front of us.

Unfortunately, everything must end. We spent our last week in Alice Springs building plywood shipping boxes which were loaded with all of the collected flora and fauna and molds we had made of various desert objects. Once labeled and fully loaded, the boxes were shipped to the ROM. Now, all that was left to do was to get ready for our long flight home. From Alice Springs to Sydney, from Sydney to Auckland, from Auckland to Hawaii, from Hawaii to Vancouver and from Vancouver back to Toronto.

It took some time before our Australian collection arrived at the museum. In the meantime we hired an outside designer to design the Australian reptile gallery. I was very pleased with this design company and found the man in charge of this outfit to be a delight to work with. He was flexible, full of good ideas and most of all he would listen to our field experiences. My idea for this exhibit was to build an open diorama portraying a dried-up Alice Springs riverbed. I thought that the visiting ROM public could walk through the dry riverbed which would be lined on both sides with the various Australian flora and fauna we had collected. I thought that this would make an informative and interesting Australian Desert display. Out of reach replicas of termite hills, spinifex and various other vegetation would line the riverbed in which many replicas of all of the species of snakes and lizards we had collected would be displayed.

As it happened, by the time the Australian collection arrived at the ROM we were told that there were no funds to build the project. After spending all of that money, sending four people to Australia, shipping and packaging all of the collected materials, I found it difficult to accept that the project was canceled. I wondered what in hell was going on in senior management managing the ROM gallery funds. However, it was what it was and I could do nothing about it. My boss, Lorne Render, had tried hard to get this gallery off the ground and gave us all of the support he possibly could. I don't think that this cancellation was his fault. The only positive outcome of all of these miscalculations was that all of the Australian collected reptiles added much to the diversity of the ROM herpetology collection.

Now that the production on all new gallery projects had been stopped, we continued to work on improving the few newly constructed projects we had managed to complete. I kept the artist and taxidermist occupied in producing work to improve, wherever we could, any of the existing life science displays. Things had, however, really slowed down and at times it was difficult to keep everybody enthusiastically involved in their work.

Meanwhile, Lorne Render and I got along very well. Whenever Lorne was away, he left me in charge of his department and we really never encountered any problems. Lorne was always supportive and generous. He pretty much let us do our jobs without any interference. Unfortunately, without any major jobs on the go and office politics running rampant, the constant bitching, the constant undermining of all I tried to accomplish finally got to me. I always tried hard to ignore these rather silly politics which at that time dominated just about all sections in the exhibit/design department. After these continuous conditions I too became negative and soon found myself not accomplishing much. Unfortunately, this caused problems between Lorne Render and me, and through no fault of Lorne's, we became disassociated. We ended up not communicating making my presence in the department rather useless. I think that this unfortunate situation was perhaps totally my fault. I had become thoroughly disenchanted in my workplace and was soon totally uninspired with all I was doing. It was during these awkward times that Lorne decided to transfer me to another department. This was a total surprise and came on a day at the ROM I will not forget. One morning, in the second week of July in 1988 Lorne called me into his office to inform me that I was being transferred to the ROM outreach department. Shortly thereafter, I was informed that my dad had suddenly past away. I remember going home after receiving all the negative news trying to figure out what sort of life lay ahead of me. I took the following day off to mentally prepare myself for my future. I knew that the outreach department had become a dumping ground for personnel no longer useful in the exhibit/design department. When I returned to work, I went to see my old friend E. J. Crossman – just to talk, to vent my dissolutions to a friendly person. As always, E. J. was sympathetic and a good listener.

Soon afterward, I cleaned out my office and tried to contact my new boss, David Young, head of the outreach department. It took several days before I finally got a hold of David Young to discuss with him my start and my responsibilities in his department. I thought, *Boy, this is one busy guy, there must be a lot of things going on in the outreach department since I found it so difficult to get a hold of him.* Finally one morning I did find David Young in his office and we discussed my new beginning. David Young turned out to be one of the nicest guys you would ever want to meet. I think he regimented his life along the Buddhist philosophy making him very easy to get along with. Unfortunately, however, David stayed at home for at least two days each week which really separated him from his staff and from the general activities of the outreach department. Everyone in the department did whatever they wanted to do. The outreach department had all the supportive equipment one could possibly ask for. It had a great departmental library, excellent office, and workspaces for all of its staff.

The outreach staff consisted of three traveling teachers, two interpreters/writers, a designer, a technician, and one person who was in charge of promoting and booking all outreach exhibits to various educational institutions throughout the province. This person also made sure that all outreach exhibits arrived at their designated destinations on time and in good condition. Each of the traveling teachers had their own car to travel to the various Ontario schools they serviced. I guess I was second in command and was responsible for all of the designated responsibilities of the department. It took some time to discover what each outreach staff member was responsible for. Once I sorted out the various responsibilities, I was ready to plan future projects not yet on the drawing board. I soon learned that for some time the department had planned to produce a traveling dinosaur exhibit. This was to be designed and installed in a full-sized trailer which was to be towed to various venues. The trailer was on a long-term loan from the Calgary Glenbow institution. This was a challenging project which, at the time, was the sort of challenge I needed. A project I could really sink my teeth into and I soon engaged all outreach staff required for the production of the dinosaur exhibit. I don't think that all of the staff was pleased in my insisting on starting to work full time on the project immediately. In retrospect, I should perhaps have left things the way they were in the department, however, this has never been in my nature to just sit around and do as little as possible.

Once the large trailer arrived at the ROM I had to find a storage area for the unit. There was no room at the museum. Eventually, I found space in the west end of Toronto in a graphic design and production company's parking lot. This company would eventually produce large murals to cover the total outside surfaces of our dinosaur trailer.

Before designing the dinosaur-traveling exhibit I thought it would be a good idea to travel to see the dinosaur museum in Drumheller, Alberta, to get some first-hand information on how dinosaurs lived and how best to exhibit them. Three of us traveled to Drumheller, including the outreach designer, the outreach exhibit promoter. Our

Drumheller excursion was an excellent idea and provided us with a lot of good information.

After returning from the west, I waited for several weeks for our designer to provide us with workable design drawings to work from. These drawings, however, never materialized. Finally I decided to design the traveling dinosaur exhibit myself. I asked the outreach technician, Bill Baker and the outreach exhibit promoter to assist me in the construction of the traveling dinosaur exhibit. Both proved to be enthusiastic and contributed much good work and some good ideas for the completion of the project.

Three of us traveled each morning to the west end of the city and worked for several weeks on the dinosaur trailer project.

Having never received any proper design drawings, we pretty much worked from my mental ideas. My final design displayed a full-size Albertosaurus assembled skeleton near the entrance of the trailer. To accomplish this, I had to hire an outside expert to cast all of the required bones and to assemble these to construct the life-sized Albertosaurus skeleton. To make this skeleton look as large as possible, I wanted to stand the skeleton some twelve inches higher than the trailer floor. Unfortunately, the person I had hired to build our dinosaur skeleton ignored my instructions and consequently mounted the assembled skeleton directly to the floor of the trailer. When I asked the man why he did not raise the Albertosaurus off the floor as he had been requested, his answer was short and rather abrupt and he replied, "I lied." Now we had to undo the large skeleton from the floor of the trailer, build a twelve-inch raised floor for the dinosaur to stand on and reinstall the Albertosaurus.

This cut into my schedule and I suggested to my boss, David Young, not to pay the outside dinosaur expert the full amount originally agreed upon. However, David ignored my suggestion and paid the outsider the full amount anyway.

We soon created the sort of habitat in which an Albertosaurus would have lived in. To accompany our simulated foreground, I painted a fair sized picture of the ancient habitat which we had blown up and mounted this enlargement to the back wall of the display. We also created some imaginary Albertosaurus sounds which added another dimension to the display. Overall, this open diorama turned out well and was appreciated by the many schoolchildren who visited the dinosaur trailer.

With the guidance of the head curator of the ROM's vertebrate paleontology department, we created several other dinosaur stories on both sides of the trailer filling all available spaces. Each of these exhibits interpreted some scientifically proven dinosaur phenomenon. For many of these exhibits, we also added sound effects, again creating that extra dimension.

Eventually, I painted a colorful picture of various species of dinosaurs which was blown up to cover the total exterior of the trailer. The pictures identified the trailers display from quite a distance. The dinosaur trailer traveled throughout Ontario, visiting small towns as well as bigger cities. While on the road, thousands of schoolchildren visited the exhibit.

When the dinosaur trailer was completed and traveled throughout Ontario, changes suddenly came about in the outreach department. The head of the department never once visited the trailer during its construction. He really showed no interest in the project. Soon after the completion of our traveling dinosaur exhibit, the head of the outreach department was dismissed from the ROM. One day, when I had returned from putting the final touches on our traveling dinosaur exhibit, I was informed by the head of the ROM human resources department that David Young had been dismissed. When I arrived at the ROM, he had already cleaned out his office and I never saw him again. It was never explained to me why he was dismissed. I guess all of his absentee days and his casual approach to all of the goings on within the department had finally caught up to him.

Shortly thereafter, I was informed that I was now in charge of the outreach department. I really did not want this job but, I was given no choice. To make things worse for me, a short time later I was informed that my former boss Lorne Render, had resigned from the exhibit design department and I was informed to combine outreach and exhibit design into one department. I was informed to make this combination more efficient. Again, I really did not want the job but I was given no choice in the matter. I would have much more enjoyed being an artist in this new department, however, like it or not, I once again ended up in a management position.

Combining the outreach and the exhibit design departments turned out to be most challenging. Senior management had decided to set up two separate departments for the production of all gallery and new temporary exhibits productions. One of these new departments employed all of the production management staff. This was run by a real office politician whose talents often made things difficult for me to manage my area of this new setup. I was put in charge of all two and three-dimensional design personnel and was to assign these as requested to the project management section. Unfortunately, I was not able to assign the designers I thought to be the best qualified for any specific project. It ended up that each production manager got to pick whomever they preferred to work with. This became very personal, and some designers, not liked by certain project managers, were left out of work.

The project management department really had the easy side of gallery production. If a designer did not do a good job, it was up to me to discipline them. This made no sense. Since the assigned designers were working directly for the project managers it was difficult for me to assess any work problems. Not only did the project management department expect me to discipline all workers they also expected me to fire any of the designers they could not get along with. The whole system did not make a hell of a lot of sense. However, that was what it was and I tried hard to make the best of what was given to me.

I had enough staff problems in my area without managing faults in the production managers' section. One of my graphic designers decided to make things even more miserable for me. I had asked this particular graphic designer to please arrive for work, like all other ROM staff members, at nine o'clock in the morning and to stay until five in the afternoon. It was her habit to drift in whenever it was convenient for her in the morning. It was resolved that since all ROM staff, particularly curatorial staff, worked from nine to five, each day, and since designers had to spend a lot of time with curators to accomplish their work, I thought it was proper for all design staff to share the same working hours. I was told to run things more efficiently. Well, after explaining my request to this particular designer I was charged with harassment. I don't remember how many meetings I had with various union people, senior management and with the head of the ROM human resources department regarding this senseless charge. Again and again I tried, in vain, to explain my request to the endless number of people becoming involved in this situation. I could not find any support from anyone in senior management. The only suggestion I got was from the head of human resources to take this specific designer out for lunch. My answer to the head of human resources was, "You take her out for lunch." After wasting a hell of a lot of time, this silly problem was never resolved.

The person put in charge of the production management area and my section was a man who used to be part of the ROM entomology department. He had ambitions to become part of ROM senior management and was now trying hard to manage the building of all new ROM exhibitions. To my surprise, one day I received a complaint from this man that I was to stop badmouthing the staff of the production management area. I had no idea what he was talking about. However, a bit later I discovered that my old friend, the bat cave designer, had gone to him and complained about me talking to the carpenters about her cave design mistakes. Years later, she was still going on about her blunders. To add to all of the silly ongoing staff problems, the senior manager in charge of the exhibit production teams did not last long and shortly after his start, he was dismissed.

By this time I got a new boss. This person was Jean Lavery – part of the senior management team. She and I got along very well. Jean was originally part of the design staff consultants who designed the ROM 'How to build museum exhibits.' Jean was a good superior. She was always there for us assisting in solving any immediate problems. She was supportive in my running the designer staff section and understood the challenges involved in supplying design staff to the production management team. Unfortunately, Jean was only in her position for a year or so. She retired and went back to her native country, England.

By the time Jean left the ROM, senior management had already hired a new person to fill her position. From day one, I did not get along with this person. Most of the staff in the project managers section was falling all over our new boss, including many of my staff. This was what she desired and those who did not suck up to her were pretty much disregarded. It was never my style to pretend and/or to relate falsehoods to my superiors. I focused on doing my job as best as I could. I always preferred a professional working relationship with my superiors. Unfortunately, this was not how things worked with the new lady in charge. She called a lot of staff meetings making sure that everyone knew who was now in charge. One of her staff meetings took place at her house. I can only guess that she wanted to impress everyone with her home. Another full staff meeting she called took place in the ROM auditorium. Shortly before this meeting, I was told by my new superior to tell one of my staff members not to attend. This was the girl who worked for me in the outreach section booking all of the traveling exhibits. This poor girl suffered from severe rheumatism. I did relate the senior management message to her, however, she decided to attend anyway. During the auditorium gathering, our new leader stood on stage, dressed in the brightest red suit I had ever seen, with both hands folded into fists and placed firmly on her hips. She then lectured us on her plans for the exhibit design sections. It soon became obvious to me that she really had no idea what a research museum was all about. Soon after that imposing meeting, I was called to her office and questioned why the above staff member she had asked not to attend her meeting did in fact attend. She asked whether or not I had informed the girl in question not to attend her meeting. I informed the boss lady that I indeed had done so. I found out later that the new boss lady, for whatever reason, wanted to fire the rheumatic girl as soon as possible. I suspected more office politics. At that time in her office, she also accused me of improper body language. To this day I do not know what the hell she was talking about. I know that I did not drop my drawers or walk around with my zipper open, nor did I point my middle finger at anyone. It was never given a reason for the accusation. Soon after this final encounter with my boss, I resigned from the Royal Ontario Museum. There simply was no way I could continue to do my job effectively. I was constantly harassed and everything I accomplished was questioned. I negotiated a buyout with the head of human resources, which would see me through for the time being, at least until I recovered from my decision to leave the ROM. I had no idea what in hell I was going to do. Having spent my whole professional life at the ROM, not having any formal education, made my future looked rather dismal. In retrospect, I should have endured the constant harassment from my superior. Within a year of leaving the museum, the lady in charge of the exhibit design processes was dismissed.

Epilogue

My life has taken many turns since my ROM experiences.

I started my own studio creating sculptures and three-dimensional exhibits. To this day, I continue to paint both wildlife and landscapes and exhibit my art in various venues. The last fish painting I did was an Atlantic salmon which I collaborated on with Bev Scott a year before his passing. It is displayed in the ichthyology department in his honor.

I have seen the death of many of my lifelong dear friends from the ROM – W. B. Scott, E. J. Crossman, Terry Shortt, and Anker Odum. How thankful I am that they were a part of my life.

The memories, friendships, and knowledge that one learns during their career stays and becomes ingrained within, shaping who we are today.